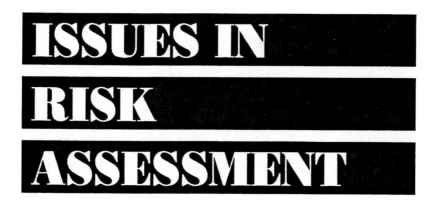

ISSUES IN RISK ASSESSMENT

Committee on Risk Assessment Methodology
Board on Environmental Studies and Toxicology
Commission on Life Sciences
National Research Council

NATIONAL ACADEMY PRESS
Washington, DC 1993

NATIONAL ACADEMY PRESS   2101 Constitution Ave., NW   Washington, DC 20418

The project was supported by the U.S. Environmental Protection Agency under Cooperative Agreement #CR-815682, the American Petroleum Institute, and the American Industrial Health Council. Funds were also contributed by the National Institute for Occupational Safety and Health and the U.S. Army Biomedical Research and Development Laboratory.

Library of Congress Catalog Card No. 92-61838
International Standard Book Number 0-309-04786-2

Additional copies of this report are available from the National Academy Press, 2101 Constitution Avenue, NW, Washington, DC 20418.

B-023

RA
427.3
N37
1993

Printed in the United States of America

First Printing, January 1993
Second Printing, March 1993
Third Printing, March 199

Committee on
Risk Assessment Methodology

BERNARD D. GOLDSTEIN *(Chairman)*, Environmental and Occupational Health Sciences Institute, Robert Wood Johnson Medical School, and Rutgers University, Piscataway, NJ

DONALD R. MATTISON *(Vice Chairman)*, University of Pittsburgh, Graduate School of Public Health, Pittsburgh

JOHN C. BAILAR, III, McGill University School of Medicine, Montreal

PAUL T. BAILEY, Mobil Oil Corporation, Princeton

LAWRENCE W. BARNTHOUSE, Oak Ridge National Laboratory, Oak Ridge, TN

KENNY S. CRUMP, Clement Associates, Inc., Ruston, LA

JOHN DOULL, University of Kansas Medical Center, Kansas City

MICHAEL A. GALLO, Robert Wood Johnson Medical School, University of Medicine and Dentistry of New Jersey, Piscataway

RICHARD A. GRIESEMER, National Institute of Environmental Health Sciences, National Toxicology Program, Research Triangle Park, NC

WILLIAM E. HALPERIN, National Institute for Occupational Safety and Health, Cincinnati

ROGENE HENDERSON, Lovelace Biomedical and Environmental Research Institute, Albuquerque

BRIAN P. LEADERER, John B. Pierce Foundation, Yale University School of Medicine, New Haven, CT

ALAN W. MAKI, Exxon Corporation, Houston

FRANKLIN E. MIRER, United Auto Workers, Detroit

DANIEL W. NEBERT, Institute of Environmental Health, University of Cincinnati Medical Center, The Kettering Laboratory, Cincinnati

D. WARNER NORTH, Decision Focus, Inc., Mountain View, CA

RICHARD H. REITZ, The Dow Chemical Company, Midland, MI

iii

Staff

RICHARD D. THOMAS, Principal Staff Scientist
GAIL CHARNLEY, Project Director
KATHLEEN R. STRATTON, Project Director (until March 1992)
MARVIN A. SCHNEIDERMAN, Senior Staff Scientist
ANNE M. SPRAGUE, Information Specialist
IAN C.T. NISBET, Technical Adviser
DANIEL KREWSKI, Technical Adviser
LINDA V. LEONARD, Senior Project Assistant
RUTH DANOFF, Project Assistant
JOYCE WALZ, Project Assistant

Sponsors

U.S. Environmental Protection Agency
American Petroleum Institute
American Industrial Health Council
National Institute for Occupational Safety and Health
U.S. Army Biomedical Research and Development Laboratory

Committee on Risk Assessment Methodology

Federal Liaison Group

WILLIAM H. FARLAND *(Co-chair)*, U.S. Environmental Protection Agency, Washington, DC

ROBERT SCHEUPLEIN *(Co-chair)*, U.S. Food and Drug Administration, Washington, DC

DEBORAH BARSOTTI, Agency for Toxic Substances and Disease Registry, Atlanta, GA

JAMES BEALL, U.S. Department of Energy, Washington, DC

JAMES BILSTAD, U.S. Food and Drug Administration, Rockville, MD

WILLIAM CIBULAS, Agency for Toxic Substances and Disease Registry, Atlanta, GA

MURRAY S. COHN, Consumer Product Safety Commission, Bethesda, MD

JOSEPH COTRUVO, U.S. Environmental Protection Agency, Washington, DC

GERALD A. FAICH, U.S. Food and Drug Administration, Rockville, MD

HENRY S. GARDNER, U.S. Army Biomedical Research Development Laboratory, Frederick, MD

HERMAN GIBB, U.S. Environmental Protection Agency, Washington, DC

WALTER H. GLINSMANN, U.S. Food and Drug Administration, Washington, DC

BRYAN D. HARDIN, National Institute for Occupational Safety and Health, Cincinnati

RONALD W. HART, National Center for Toxicological Research, Jefferson, AR

RICHARD N. HILL, U.S. Environmental Protection Agency, Washington, DC

Board on Environmental Studies and Toxicology

Commission on Life Sciences

Preface

This volume is the first in a series to be prepared by the Committee on Risk Assessment Methodology (CRAM) in the National Research Council's Board on Environmental Studies and Toxicology. The committee was charged with identifying and investigating important scientific issues in risk assessment. Three issues related to risk assessment are addressed here: use of the maximum tolerated dose (MTD) in animal bioassays for carcinogenicity, the two-state model of carcinogenesis, and a paradigm for ecologic risk assessment.

The use of the MTD in animal bioassays has been standard practice in the United States for more than 15 years, and controversy surrounding its use is not new. However, questions continue to be raised about the utility of the data derived from such tests and about the validity of inferences drawn from the data. Stimulated by the informaiton presented in a workshop held on September 6, 1990, and discussions held at later meetings, CRAM has examined the issues related to the MTD. The first report in this volume contains its findings and recommendations on the issues. The workshop included presentations by Eugene McConnell on "Definition and Application of MTD," by Daniel Krewski on "Correction Between the MTD and Measures of Carcinogenic Potency: Implications for Risk Assessment," and by Bruce Ames on "What Are Bioassays Conducted at the MTD Telling Us?" The program, a workshop summary, and a list of attendees appear as appendixes to the first report in this volume. Dr. Krewski's presentation summarized findings from a review paper with the same title, which was developed specifically for the workshop. Invited to discuss the presentations were Edmund Crouch, Kenny Crump, John Emmerson, Reto Engler, Michael Gallo, David Gaylor, Ian Munro, Thomas Starr, James Wilson, and Lauren Zeise.

In the second report in this volume, CRAM examines the use of the two-stage model of carcinogenesis, which is based on a paradigm that is thought to reflect the biologic mechanisms underlying carcinogenesis, for human risk assessment. Like the use of the MTD, the use of empirically based mathematical models for evaluating the relationship between dose and response in rodent bioassays and extrapolating from high to low doses is standard. However, questions have been raised about the biologic relevance of such procedures and about the validity of human risk assessments based on the models. This report was based on information presented in a workshop held on November 8, 1990, and discussions held at later meetings. The workshop included presentations by Alfred Knudson on "Biological Factors in Two-Stage Models," by Suresh Moolgavkar on "Two-Stage Clonal Expansion Model of Carcinogenesis," and by Samuel Cohen on "Application of the Two-Stage Model to Animal Data." Invited to discuss those presentations were Carl Barrett, William Farland, Robert Maronpot, Robert Sielken, Todd Thorslund, and James Wilson.

The third report in this volume examines the overall process of ecological risk assessment and was stimulated by information presented at a workshop held on February 26-March 1, 1991, and discussions held at later meetings. The workshop included numerous speakers and discussants, whose goals were to survey existing approaches to ecological risk assessment, consider developing a consistent framework for ecological risk assessment, and identify major uncertainties and research needs. The keynote speakers were Terry Yosie, of the American Petroleum Institute; Michael Slimak, deputy director of the Office of Ecological Processes and Effects Research, U.S. Environmental Protection Agency; and Warner North, of Decision Focus, Inc., a member of the committee.

Some of the other reports being prepared by CRAM will re-evaluate established practices or principles in light of potential alternatives, and some will address new concepts to advance the science of risk assessment. It is hoped that the series of reports that result from the committee's deliberations will help scientists in regulatory agencies, academe, and industry to find common ground for defining, understanding, and discussing important ideas in the field.

The conclusions and recommendations presented herein were arrived at by the committee in executive session. Thus, the scientific interpretations are those of committee members and not necessarily those of other participants in the workshops. The committee's reports were reviewed according to standard NRC practices, and the committee thanks the reviewers for their close attention and useful comments. The workshop summaries in the appendixes were prepared as working papers for the committee by the workshop organizers and participants; they are not NRC reports and have not been subjected to NRC review procedures.

The committee thanks the persons who participated in the workshops, especially the speakers, whose presentations provided important information for the consideration of the committee. Special thanks also are given to the members of the federal liaison group, whose names and affiliations are listed in the front of this report.

Two task groups of the committee took special responsibility for the workshops and reports. Although the entire committee shares the responsibility for the contents of the reports, the task-group members listed below must be credited for having done the key work of organizing the workshops and preparing their findings and recommendations for review and endorsement by the full committee.

No effort of this kind can be accomplished without the hard work and dedication of a talented staff. The committee joins me in thanking the following staff of the Board on Environmental Studies and Toxicology: James Reisa, Richard Thomas, Gail Charnley, Kathleen Stratton, Mary Paxton, Marvin Schneiderman, Anne Sprague, Ruth Danoff, and Linda Leonard.

Bernard Goldstein
Chairman, CRAM

MTD and Two-Stage Model Task Group Members

Kenny S. Crump, Chairman, MTD
Richard A. Griesemer, Chairman, Two-Stage
Paul T. Bailey
Michael A. Gallo
Rogene Henderson
Donald R. Mattison
Richard H. Reitz

Ecological Risk Assessment Task Group Members

Lawrence W. Barnthouse, Chairman
Alan W. Maki
D. Warner North

Technical Advisers

Daniel Krewski
Lois Gold
Ian C.T. Nisbet

Contents

THE TWO-STAGE MODEL
OF CARCINOGENESIS

Issues in Risk Assessment

Executive Summary

Risk assessment is a relatively new and rapidly developing science. Indeed, most federal agencies for which risk assessment is an important tool for decision-making or a subject of research were established only within the last quarter-century. Among those are the Environmental Protection Agency (EPA), Occupational Safety and Health Administration (OSHA), Consumer Product Safety Commission (CPSC), National Institute of Environmental Health Sciences (NIEHS), National Institute for Occupational Safety and Health (NIOSH), Food and Drug Administration (FDA), and Agency for Toxic Substances and Disease Registry (ATSDR). Mantel and Bryan published in 1961 the first paper on estimation of low-dose risk based on data obtained from tests in which animals were exposed at high doses; formal procedures for performing animal bioassays, which are critically important for gathering information for risk assessment, had been standardized only in the 1960s and 1970s; and formal risk assessment began to be conducted regularly in the late 1970s. It was not until 1983, when the National Research Council (NRC) committee that prepared *Risk Assessment in the Federal Government: Managing the Process* defined the steps in risk assessment, that a generally accepted nomenclature for risk assessment was established.

Now, after this short time, risk-assessment scientists study the details and argue the relative merits of different approaches to the performance and interpretation of studies; learned societies publish journals to communicate these deliberations; and national and international meetings are convened to discuss specific issues or to write the blueprints for new programs. New concepts are being rapidly explored, such as the use of pharmacokinetic studies of the fate of a chemical agent in the body; and some of the practices and principles established only a few years ago are already being re-evaluated. In addition, whereas almost all efforts were once directed toward determining the carcinogenic potential of an agent, scientists are now equally interested in assessing the potential of *mixtures* of agents to produce not only cancer, but reproductive, neurotoxic, developmental, and immunologic effects.

This volume contains the first three reports the Committee on Risk Assessment Methodology (CRAM) in the National Research Council (NRC) Board on Environmental Studies and Toxicology. The committee's work was sponsored by a consortium of federal agencies and private organizations, including EPA, NIOSH, the U.S. Army Biomedical Research and Development Laboratory, the American Petroleum Institute, and the American Industrial Health Council. The committee was charged to assess the scientific basis, inference assumptions, and regulatory uses of and research need in risk assessment. The committee has investigated these issues partly through a series of narrowly focused workshops. Topics were chosen in consultation with federal regulatory agencies on the basis of scientific considerations and the needs of the agencies. One source is a list of subjects that appeared in *Risk Assessment if the Federal Government*, which has become known as the Red Book. CRAM's reports are intended to provide guidance to regulatory decision-makers on specific questions; they are not broad, thorough scientific analyses, as are many NRC reports. The committee has focused on methodology; accordingly, its deliberations on each topic takes into account not only potential problems with existing methods, but also the suitability of alternative methods for risk assessment.

The committee consulted closely with federal agencies whose mission is to make decisions based on risk assessment of environmental and human health hazards. Representatives of 11 federal agencies organized themselves as a federal liaison group, and the committee consulted with the group in selecting workshop topics and participants and in preparing workshop summaries. However, in accordance with NRC policy, the members of the federal liaison group did not take part in the committee's deliberations or in the preparation of its reports. The workshop presentations, commissioned papers, and extensive committee deliberations formed the basis for the findings in the reports.

The committee began meeting in January 1990 and selected as its first topic of study and use of the maximum tolerated dose (MTD) in animal bioassays, with emphasis on the relationship between the MTD and the carcinogenic potency of a test chemical. The second topic was the two-stage model of carcinogenesis, with a focus on data requirements for regulatory application. The third topic was a conceptual framework for ecologic risk assessment. The committee's reports on those three subjects make up this volume. Two other topics that have been selected are exposure assessment and developmental toxicity; workshops on these topics have been held, and reports are in preparation.

Use of the Maximum Tolerated Dose in Animal Bioassays for Carcinogenicity

Long-term animal bioassays for carcinogenicity are used regularly to determine whether chemical agents are capable of inducing cancer in exposed animals. Two important aspects of current bioassays are that testing covers a substantial portion of the lifespan of the test species and that high doses are used. The highest dose tested (HDT) is an approximation of the maximum tolerated dose (MTD), which is roughly described as the highest dose that does not alter the test animal's longevity or well-being because of noncancer effects.

The committee chose as its first task to address the use and limitations of MTD testing in long-term animal bioassays for carcinogenicity. The first report focuses specifically on whether the MTD should continue to be used in carcinogenicity bioassays, and it does not address all the issues related to performing carcinogenicity bioassay or interpreting their results.

In particular, the committee chose to investigate the observation that statistical analyses of the results of bioassays of many chemicals have shown strong correlations between measures of carcinogenic potency, such as the TD_{50} (the dose that causes tumors in 50% of test animals that would otherwise be tumor-free), and measures of toxicity, including the MTD. The strength of the correlations suggests that carcinogenicity is inherently related in some way to other toxic effects produced by a chemical, although dependence on such factors as the bioassay design and the mathematical and statistical methods used to estimate potency and investigate the correlations has also been proposed.

The committee concluded that the correlations are not wholly mathematical or statistical artifacts, but are due partially to an underlying relationship between measures of general toxicity (e.g., the MTD) and measures of carcinogenic potency. The relationship can be expressed as follows: increases in cancer incidence large enough to be detected (i.e., to be statistically significant) in standard bioassays generally occur only at doses near the MTD. The committee suggests that because of the relationship between TD_{50}s and the MTDs, a preliminary (and perhaps uncertain) estimate of the potential carcinogenic potency of an untested chemical can be derived from its MTD. Such an estimate is a plausible upper bound on the carcinogenic potency of a chemical, if in fact it is a carcinogen. Such estimates can prove useful in setting priorities for carcinogenicity testing and in estimating cancer risk when carcinogenicity data are not available. If an upper-bound estimate predicts a small human risk, a chemical could be given a low priority for carcinogenicity testing or might be deemed suitable for use with less extensive testing than might otherwise be required.

The committee noted that because specific criteria for selecting the HDT vary, even under the current guidelines, reports of bioassay results should include a clearly stated rationale for dose selection and a summary of the toxicity information important for evaluating the dose selection to facilitate interpretation.

The usefulness of information from bioassays conducted at the MTD has been questioned for several reasons. First, some believe that the proportion of compounds found to be carcinogenic at the MTD is so large that regulatory attention and public concern might be applies to agents that pose only trivial hazards. (The committee did not review such regulatory attention.) Second, it has been argued that some agents induce cancer at the MTD through mechanisms that do not occur at lower doses. Several mechanisms of carcinogenesis have been hypothesized to be effective only at high doses, such as increased cell proliferation rates in response to high-dose toxicity or as a result of receptor complex-mediated alterations in cell-growth control. According to these hypotheses, exposure at lower doses, where these mechanisms are inactive, would not result in a carcinogenic response. (The committee noted several examples of agents for which these hypotheses had been proposed, but did not reach conclusions on their proof or consensus on the generality of their application.) Third, it has been asserted that current bioassays, which generally involve only doses at or near the MTD, provide little information that is useful for defining the dose-response relationship. Defining the shape of the dose-response curve at lower doses would provide information that has greater relevance to human exposures and consequent risks. (The committee noted that validation of methods for extrapolation of dose-response relationship over wide ranges was beyond the scope of the study, although human exposure to some carcinogens at doses approximating those used in bioassays is known to occur.)

The committee noted several limitations in the information provided by current bioassays that use the MTD. Those assays often do not incorporate doses smaller than one-fourth of the MTD, so they do not provide direct information on the carcinogenic potential of a test substance at lower doses. But tests conducted at lower doses will probably have little power to detect carcinogenic effects, unless the number of animals tested is increased immensely, which would increase the cost of a bioassay commensurately; the large number of animals required for detection of the smaller increase in tumors incidence that might occur at low doses is one of the primary reasons for use of the MTD in carcinogenicity bioassays. Testing at doses that induce overt toxicity, however, can lead to changes in an animal's food consumption, recurrent cytotoxicity, and hormonal imbalance, all of which an increase or decrease carcinogenic responses at particular target sites. A rodent bioassay might yield information whether a chemical produces tumors in rodents, but generally can provide only scanty information on whether it produces tumors through generalized indirect mechanisms or directly as a result of its specific properties. Other data are required for extrapolating bioassay results to other doses or from animals to humans or for evaluating the possibility that indirect mechanisms of carcinogenesis can contribute to the results.

Despite those limitations, the majority of the committee concluded that current bioassays that incorporate the MTD provide some information that is useful for hazard identification and risk assessment. The assays identify substances that do or do not increase the incidence of cancer under standardized test conditions; in the case of substances that do not increase the incidence, the assays provide an operational definition of *noncarcinogen*. They identify target organs that show which tumor types are associated with exposure, thereby providing guidance for epidemiologic studies, although concordance among species is often absent. They also provide a basis for interspecies comparisons and they provide useful information on the carcinogenic potency of a chemical at high doses, as well as on differences in sensitivity between the sexes and among different strains and species of rodents, which are the test animals almost universally used.

The committee recognizes that bioassays conducted at the MTD are not designed to provide information on a biochemical and physiologic mechanisms of tumors production. Nor do they provide direct information on the shape of the dose-response curve at doses below the lowest experimental dose, which often include doses to which humans are exposed.

The committee considered four major options for modification of current bioassay procedures: (1) retain the status quo, possibly with the addition of doses lower than the MTD; (2) use a high dose that is an arbitrary fraction of the estimated maximum tolerated dose; (3) redefine the MTD, basing it on studies of the dose dependence of physiologic effects expected to alter carcinogenic response; and (4) use MTD testing as part of an overall testing strategy that separates carcinogens from noncarcinogens but also provides additional information useful for determining human relevance.

After extensive deliberation and consideration of those options, the committee was unable to come to a unanimous decision on a recommendation. Two points of view emerged. The majority of the committee considered option 4 (which recommends that the MTD, as currently defined, continue to be one of the doses used in carcinogenicity bioassays) to be appropriate and prudent. However, a sizable minority (six of the 17 committee members) did not fully agree with the conclusions and recommendations reached by the majority and prepared an alternative recommendation. The two groups' recommendations are summarized below.

The majority of the committee prefers option 4 and recommends that the MTD, as currently defined, continue to be one of the doses used in carcinogenicity bioassays. Other doses, from one-half to possibly one-tenth of the MTD or even smaller, should also be used, taking into account the capacity of the test animals to metabolize the test substance. If bioassay results are negative in both sexes of two species, generally no additional tests related to carcinogenicity are required. If the results are positive, additional studies should be performed to reduce uncertainties in the prediction of human responses to the material and in the quantification of human risk. The additional studies should address mechanisms of cancer induction, toxicokinetics and metabolism of the substance, and physiologic responses induced by the substance. The committee notes that regulation of a chemical can be instituted (for public-health reasons and to protect human lives while more data are being collected) at almost any stage of data collection and that regulation can be modified as additional data become available.

The minority of the committee believes that the process for selecting doses to be used in a carcinogenicity bioassay should be modified (option 3). Specifically, the minority recommends that dose selection be done by a panel of experts on the basis of careful evaluation of appropriate subchronic studies conducted before the bioassay is initiated. The HDT should be chosen as the highest dose that can be expected to yield results relevant to humans, not simply the highest dose that can be administered to animals without causing early mortality from causes other than cancer (i.e., the MTD as currently defined). (In contrast, the majority believes that the decision regarding results obtained with the MTD can best be made after the MTD data are collected and that future decisions—regarding either regulation or additional studies—are better grounded if these data are present than if they are absent. The minority recognizes that chronic animal bioassays were originally designed to answer a simple question: Can chemicals cause cancer in animals? It is clear that the primary motivation for conducting the chronic bioassay today, however, is to determine whether the substances tested are likely to pose a substantial cancer risk to human populations. Therefore, the minority finds that a core of basic information should be gathered before the chronic bioassay is initiated, so that the study can be designed to achieve its objective.

The minority therefore recommends that the HDT in a bioassay be selected after a careful evaluation of results of subchronic studies conducted before the 2-year bioassay (option 3). In option 3, a core of basic information gathered before the bioassay would include information about the mechanisms of toxicity in test animals and an elucidation of the dose-response curve for such toxicity. That information is important because there is concern that induction of substantial toxicity throughout the lifetime of an animal might affect the rate at which tumors develop. Information would also be required on how dosage (including repeated exposures) affects biochemical and physiologic processes that are responsible for homeostasis, cell proliferation, hormonal balance, and the uptake and metabolism of the test chemical. All those processes are known to influence cancer incidence.

In some circumstances, adoption of option 3 would not change the magnitude of the HDT. For example, if human populations were exposed to high concentrations of the test substance, the HDT might be the MTD. However, in many cases, the HDT could be much lower than the current MDT, and the range of doses tested might be much wider than that used in current studies.

The principles recommended by Sontag et al. in 1976 (and endorsed by the majority of the committee) were designed to minimize the frequency of false-negative results (i.e., to maximize the sensitivity of the bioassay). The minority believes that the changes it recommends would improve the relevance of the bioassay for human populations by increasing the specificity of the test. (The majority points out that any increase in specificity resulting from the change proposed by the minority would be accompanied by a decrease in sensitivity, and the committee did not investigate the extent to which the change would allow human carcinogenicity to go undetected.)

The minority recognizes the implementation of option 3 would not be trivial. Guidelines for the amount of information required before bioassays are initiated would have to be modified. Criteria for dose selection would vary from chemical to chemical. It is clearly beyond the scope of the minority recommendation to specify all the details for this paradigm shift. However, the minority believes that implementation of its recommendation is feasible within the current testing framework. For example, review of scientific criteria for selection of bioassay doses by the National Toxicology Program (NTP) could be carried out by its Board on Scientific Counselors. (The board currently reviews the selection of compounds to be tested by NTP and reviews NTP's reports, but does not review the selection of doses for testing by NTP.) Other testing organizations might use other review boards before commencing studies.

The inability of the committee to come to unanimity on its primary recommendations reflects differing judgments on which approach to carcinogenicity testing would be most effective in providing information to assist risk managers, given the incomplete scientific understanding of chemical carcinogenesis in rodents and humans.

The Two-Stage Model of Carcinogenesis

Efforts to improve cancer risk assessment have resulted in the development of a mathematical dose-response model, called the two-stage model, that is based on a two-stage paradigm for the biologic phenomena thought to be associated with carcinogenesis. This paradigm is based on the relationship between tumor incidence and age, which suggests that at least two critical cellular changes are necessary for the development of many nonhereditary tumors. Current evidence suggests that some tumors might require more than two critical events to be expressed as human cancer. More complex models might be needed to describe multistage carcinogenesis accurately; however, it is hoped that the two-stage model will provide more accurate estimates of the cancer potency of chemicals that the multistage models currently in use by regulatory agencies.

Applying the two-stage model requires more extensive biologic data than current procedures; and because its feasibility as a tool for routine regulatory use has been questioned CRAM chose as its second task to evaluate the data needs and regulatory applicability of two-stage models of carcinogenesis. The committee considered several applications of the two-stage model to rodent carcinogens with different mechanisms of action and different quantities of available data. The committee noted that numerous assumptions were required to apply the model in each case. Assumptions must be made about mechanisms of action, appropriate target cells, time dependence, and the shape of the dose-response relationship. Extensive data would have to be obtained to reduce the current uncertainty in these assumptions. In fact, for very few chemicals are data sufficient to support the use of this model.

By studying specific application of the two-stage model, the committee determined that when different forms of the model are consistent with a particular data set, risk estimates can differ by several orders of magnitude. Therefore, the committee concluded that even if an agent's

mechanisms of action are well understood, it will be still be very difficult to determine its dose-response relationship accurately enough to
predict doses that correspond to risk as low as one in a million.

The strength of the two-stage model for application in cancer risk
assessment is its ability to use information about intermediate steps in
carcinogenesis; however, it is difficult to characterize these steps. Few
experimental data sets now available provide all the types of data required. Before the two-stage model can be adopted for routine health
risk assessments, it will be necessary to expand current rodent bioassay
methods so that the necessary data are generated. The two-stage model
can be used now to gain insights into induced carcinogenesis, such as
identifying and characterizing the critical events, as well as to examine
the ranges of assumptions. The committee strongly encourages further
development and continued applications of the two-stage model to gain
insight into its usefulness.

Two general approaches have been used for fitting two-stage models
to data. One involves specifying trial values of parameters and simulating the subsequent tumor response. Values are then varied until the
realizations conform to the data. The second approach involves applying
standard statistical data-fitting methods (e.g., the methods of maximum
likelihood). The former approach can be quite useful in some circumstances (such as exploratory data analysis). However, the committee
encourages the use of formal statistical methods, whenever possible, to
estimate values of parameters, assess goodness of fit, calculate statistical
confidence intervals for values of parameter sand risk estimates, and
determine the extent to which the model is consistent with other mathematical representations and ranges of risk.

Two-stage models can be used as a basis for decision-making if there
is sufficient mechanistic understanding and a sufficient data base for the
chemical in question. At present, it is recommended that the two-stage
model be used primarily to increase understanding. For health risk
assessments, the two-stage model can be used with other models to add
perspective and scope to the evaluation.

A Paradigm for Ecologic Risk Assessment

The third issue addressed by the committee and the subject of the last report in this volume is a conceptual framework for ecologic risk assessment, defined as the characterization of the adverse ecologic effects of environmental exposures to hazards imposed by human activities. The workshop held on this subject had three principal goals: (1) to survey existing approaches to ecologic risk assessment through case studies representing various types of environmental stresses, (2) to consider the feasibility of developing a consistent framework for ecologic risk assessment analogous to the four-part health risk assessment framework proposed in the 1983 NRC report, and (3) to identify major scientific uncertainties and research needs common to many types of ecologic risk assessments.

The committee's principal conclusion is that, despite the diversity of subject matter and approaches taken in many different studies of ecologic stresses, a conceptual framework similar in form to that of the 1983 framework is applicable to ecologic risk assessments. However, for general applicability to ecologic assessments, the 1983 scheme requires augmentation to address some common grounds between science and management, primarily because of the need to focus on appropriate questions relevant to applicable environmental law and policy under different circumstances. Specifically, the scheme needs to address the influence of legal and regulatory considerations on the initial stages of ecologic risk assessment and the importance of characterizing ecologic risks in terms that are intelligible to risk managers. The committee's opinion is that such augmentation is as important for human health risk assessment as it is for ecologic risk assessment.

Although ecologic risk assessment and human health risk assessment differ substantially in their scientific disciplines and technical problems, the committee believes that the underlying decision process is the same for both. Therefore, the committee recommends that a uniform framework be adopted and applied to ecologic and human health risk assessment—a framework that is flexible and able to facilitate communication between scientists and risk managers. The committee extends the 1983 framework to satisfy those requirements.

The committee recommends that the use of risk assessment in strategic planning and priority-setting be expanded so that financial resources of state and federal environmental agencies can be focused on critical environmental problems and uncertainties.

The committee also recommends that research programs be established and maintained to improve the credibility of ecologic risk assessments and that ecologic risk assessments be followed by systematic research and monitoring to determine the accuracy of their predictions and to resolve remaining uncertainties.

The committee identified the following kinds of research as likely to provide major opportunities for advancement of ecologic risk assessment:

- Extrapolation across scales of time, space, and ecologic organization.
- Quantification of uncertainty.
- Validation of predictive tools.
- Valuation, especially quantification of "nonuse" values (values for environmental attributes that cannot be bought or sold).

Finally, the committee recommends that expert committees drawn from the academic community, the private sector, and regulatory agencies develop technical guidance on the scientific conduct of ecologic risk assessments.

Issues in Risk Assessment

Use of the Maximum Tolerated Dose in Animal Bioassays for Carcinogenicity

1

Introduction

BACKGROUND

The long-term animal bioassay for carcinogenicity was developed during the 1960s and early 1970s primarily as a *qualitative* screen for carcinogenic potential. Long-term animal bioassays are now used regularly to determine whether chemical agents are capable of inducing cancer in exposed animals. The bioassays are also commonly used as a basis for making qualitative inferences about the likelihood that an agent poses a carcinogenic hazard for humans as well (IARC, 1991).

Because of practical considerations, such as the cost of maintaining large numbers of animals for long periods, the number of animals used in long-term studies is generally limited to about 50 per dose-sex-species group tested. That limits the sensitivity of the carcinogenicity bioassay: it cannot detect a small increase in tumor incidence, such as an increase of 1% or less, even in experiments that use hundreds of animals. To minimize the number of false-negative results, the bioassay design was modified early in its development. The most important modifications were extension of the testing period to cover most of the lifetime of the experimental animals (which, for practical reasons, limited the test species to small rodents with lifetimes of 2-3 years) and the use of high doses.

A carcinogenicity bioassay generally involves animals exposed at two or more doses and a control group. A higher dose generally is more likely than a lower dose to produce cancer in the test animals and hence to increase the likelihood that a carcinogen will be detected. However, too high a dose might cause toxic effects that shorten the life of the test animals and prevent the observation of an excess tumor incidence. Those considerations led to the practice of selecting the maximum tolerated dose (MTD) as the highest dose tested (HDT) in an animal bioassay. The MTD is roughly described as the highest dose that does not alter the animals' longevity or well-being because of noncancer effects (Sontag et al., 1976; McConnell, 1989). These terms are further defined later in the report.

The MTD is generally estimated in a preliminary study by subjecting small groups of animals to a series of doses (perhaps six) for a small fraction of a lifetime (e.g., 3 months for mice and rats). The highest dose judged to cause no overt toxicity and little or no growth suppression is the estimated maximum tolerated dose (EMTD).[1]

Estimation sometimes results in selection of an EMTD that is too high—that causes animals to die early in life before chemically induced cancers could occur. Because it is difficult to interpret the results of animal bioassays when animals die prematurely, the bioassay design was refined to include testing at a lower dose as well—often half the EMTD (EMTD/2). Other doses (such as EMTD/5 or EMTD/10) are also used to define dose-response relationships better. Current bioassay designs have become reasonably well standardized and usually specify lifetime testing of both sexes of two species of rodents at two or more doses, the highest of which is the MTD (IARC, 1986a,b). Criteria for interpreting results obtained in these tests and for classifying them as positive, equivocal, or negative have been developed and refined (see the technical reports series of the U.S. National Toxicology Program and many others).

Although the carcinogenicity bioassay in rodents was developed primarily for *qualitative* screening of agents for carcinogenicity, it often provides the only *quantitative* information for evaluating the relationship between dose and carcinogenic response and for estimating the carcinogenic potency of an agent. Procedures for quantitative risk assessment

---

[1] MTD will be used throughout this report, except where precision requires the distinction between MTD, EMTD, and HDT. A bioassay that uses an EMTD as its HDT will be referred to as an "MTD bioassay."

were developed, beginning mainly in the 1970s, to meet the needs of regulatory agencies charged with developing reasonable limits on human exposure to agents that had been identified as potential carcinogens (Mantel and Bryan, 1961). These procedures use mathematical models and supplementary information to extrapolate data obtained from high-dose animal tests to quantitative assessments of risks to humans who might be exposed to much lower doses. Numerous risk assessments by federal regulatory agencies have been based on animal carcinogenicity bioassays.

Since 1970, several hundred chemical agents have been tested for carcinogenicity in bioassays of standard designs. The National Toxicology Program (NTP) alone has reported on 382 bioassays, of which 195 (51%) identified the tested chemical as carcinogenic under the conditions of the bioassay in at least one species-sex group (R. Griesemer, NIEHS, pers. comm., 1991). That proportion is not representative of chemicals in general, however, because of how the chemicals were selected for testing. Most of the substances (255 of 382, or 67%) were selected for testing primarily because of suspicion of carcinogenicity, and 169 (66%) of the 255 were positive. The remainder (127 of 382, or 33%) was selected for testing mainly on the basis of human exposures and the lack of toxicity data, and only 26 (20%) of the 127 were positive (R. Griesemer, NIEHS, pers. comm., 1991).

Limitations inherent in using the MTD approach and suggestions for improvement have been the subject of controversy since its use became standard (Shubik, 1978). In recent years, the use of data from bioassays performed with the MTD has been called into further question. Some of the criticisms of such data are based on the following points:

• A large percentage of chemicals tested by the NTP have been identified as carcinogenic in at least one species-sex group. Some observers believe that the test is labeling so many substances as carcinogenic that regulatory attention and public concern have been focused on many agents that pose only trivial hazards, while attention has been diverted from other agents that pose more important carcinogenic risks. The committee was given some evidence to support that charge. However, a high proportion of materials found positive in one or more species-sex groups have not been regulated (OTA, 1987).

• At high doses (including MTD and MTD/2), some agents might induce cancer through mechanisms that do not occur at lower doses, thereby generating false-positive inferences of hazard and risk for humans who are exposed at lower doses. The committee was given pharmacokinetic and other mechanistic arguments, such as induced cell proliferation, that support this hypothesis.

• Even in cases where effects might occur as a result of low-dose exposure, the results of an MTD test might have little utility in defining the dose-response relationship. Some agents could have nonlinear dose-response relationships that reflect pharmacokinetics, induced cell proliferation, or other mechanisms. The result of the nonlinearity could be overestimation (or, in some cases, underestimation) of low-dose risks. Overestimation could occur where the dose-response curve has a shallow slope at low doses and becomes markedly steeper at higher doses. Underestimation could occur where the dose-response curve flattens out or curves downward at high doses.

• Statistical analysis of bioassay results for many agents has shown strong correlations between estimates of carcinogenic potency and measures of toxicity (including the MTD) that suggest that carcinogenicity is inherently related in some way to other toxic effects produced by a chemical. However, some investigators have concluded that those correlations, and possibly estimates of carcinogenic potency, are determined in some way by the bioassay design or the mathematical and statistical methods used to estimate potency and investigate the correlations, rather than by inherent biologic properties of the agents.

SCOPE OF REPORT

The above points are all addressed in various degrees in this report. Particular attention is focused (in Chapter 2) on the fourth point—questions concerning the observed correlations between measures of carcinogenic potency and the MTD. The report explores the extent to which the correlations appear to reflect some underlying biologic reality, as opposed to being determined solely by experimental design or statistical methods. It further considers the relationship of the correlations to possible biologic mechanisms of carcinogenesis and the implications of the correlations for risk assessment.

The report discusses both what bioassays conducted at the MTD can tell us and what they cannot tell us, qualitatively and quantitatively, regarding carcinogenic hazard in humans (Chapter 3). Several proposals are discussed (Chapter 4) for modifying the design of the bioassay, for modifying the process of selecting chemicals for testing, and for augmenting the results of the bioassay with additional testing to improve risk assessments.

The committee's conclusions are presented in Chapter 5 with the recommendations of the majority of the committee concerning the better use of bioassays, specific results from bioassays, and other types of data to assess carcinogenic hazards in human populations. The dissenting recommendations of a minority of the committee are also described.

An active discussion is in progress in the scientific community concerning the extent to which high doses produce increased mitogenesis (cell division) and how much the increase contributes to the incidence of cancer at the MTD and lower doses. That was the subject of a presentation given at the MTD workshop conducted by the committee (see the summary of the workshop at the end of the report). This report reviews and evaluates the recent research related to the issue.

Although the MTD concept is used in other contexts (e.g., tests for reproductive toxicity and teratogenicity), the committee essentially limited its investigation to the use of the MTD in bioassays for carcinogenicity associated with exposure to chemicals.

2

Correlations Between Carcinogenic Potency and Other Measures of Toxicity

DEFINITIONS AND BACKGROUND

McConnell (1989) has provided a definition of the maximum tolerated dose (MTD) and explained how it is determined as follows:

> Sontag et al. (1976) had defined the MTD as "the highest dose of the test agent during the chronic study that can be predicted not to alter the animals' longevity [through] effects other than carcinogenicity" and stated that it should cause "no more than a 10% weight decrement, as compared to the appropriate control groups, and . . . not produce mortality, clinical signs of toxicity, or pathologic lesions (other than those that may be related to a neoplastic response) that would be predicted to shorten an animal's natural life span." That definition has been modified, but is still essentially the same. However, the main characteristic now used in selecting the MTD is histopathologic appearance; weight is a secondary consideration.

The estimated maximum tolerated dose (EMTD) is based on a 90-day or other subchronic test, and its determination involves scientific judgment applied to the information available at the end of the test period. How well the EMTD approximates the true MTD can be evaluated only after the bioassay. The highest dose-tested HDT in a long-term rodent bioassay is usually used as the EMTD. However, that was not always the case, especially in bioassays conducted before testing at the EMTD became the standard practice.

The MTD, by definition, is an inverse measure of the potency of an agent in causing chronic toxic effects, specifically those manifested as premature death, weight loss, or histopathologic changes after near-lifetime exposure. Potency refers to the range of doses over which a chemical produces increasing responses. Chemical A is considered more potent than chemical B if more of B than of A is required to elicit an identical response.

The LD_{50} (dose that is lethal to 50% of animals tested) is an inverse measure of the acute toxicity of an agent. It is defined as the dose (in milligrams per kilogram of body weight) that is expected to kill half a set of animals after a single administration.

The TD_{50} is an inverse measure of the carcinogenic potency of an agent and was defined by Peto et al. (1984) as follows:

> For any particular sex, strain, species and set of experimental conditions, the TD_{50} is the dose rate (in mg/kg body weight/day) that, if administered chronically for a standard period —the "standard lifespan" of the species—will halve the mortality-corrected estimate of the probability of remaining tumorless throughout that period.

Gold et al. (1984, 1986a,b,c, 1987a, 1989a,b, 1990) have tabulated estimates of the TD_{50} for individual tumor sites and (in some cases) total tumors from more than 4,000 sets of tumor data on 1,050 chemicals. The criteria used by Gold et al. (1984) in deciding what chemicals to include in their Cancer Potency Data Base (CPDB) were as follows:

A.   National Cancer Institute (NCI)/NTP bioassay, or
B.   Bioassay in the published literature meeting all the following criteria:

• Animals tested were mammals,
• Administration was begun early in life (100 days of age or less for rats, mice, and hamsters),
• Route of administration was diet, water, gavage, inhalation, or intravenous or intraperitoneal injection (i.e., where the whole body was

more likely to have been exposed than only a specific site, as with sub-cutaneous injection or skin painting),

• Test agent was administered alone, rather than in combination with other chemicals,

• Exposure was chronic, with not more than 7 days between adminis-trations,

• Duration of exposure was at least one-fourth the standard lifespan of the test species,

• Duration of experiment was at least half the standard lifespan of the test species,

• Research design included a control group,

• Research design included at least five animals per group,

• Surgical intervention was not performed,

• Pathology data were reported as the number of animals with tu-mors, rather than the total number of tumors,

• Results reported were original data, rather than secondary analyses of experiments already reported by other authors.

Bioassays of particulate or fibrous matter and of mixtures of chemicals were not included (except some commercial preparations to which hu-mans are often exposed).

The CPDB or the computerized National Toxicology Program/ National Cancer Institute (NTP/NCI) database served as the data sources for the statistical analyses of correlations between carcinogenic potency and other measures of toxicity conducted by a number of investigators (Zeise et al., 1984, 1985, 1986; Bernstein et al., 1985; Crouch et al., 1987, and Rieth and Starr, 1989a,b). In particular, in a paper specifi-cally prepared for the present committee's workshop on the MTD, Krewski et al. (Appendix F) calculated estimates of the TD_{50} for a sub-set of 191 chemicals listed in the CPDB; they used three models of the dose-response relationship: the single-stage models used by Peto et al. (1984), a multistage model, and a Weibull (in dose) model.

In addition to the TD_{50}, carcinogenic potency can be measured on the basis of the slope of the dose-response curve in the low-dose region, expressed by the parameter q_1. The parameter q_1 is the coefficient of the linear term in the multistage model of Armitage and Doll (1961) as adapted for risk assessment by Crump (1984). When the model is ap-plied to experimental data on tumor frequencies, q_1 is an estimate of the carcinogenic potency of an agent at low doses. The statistical upper confidence limit on q_1, denoted q_1^*, can be determined by the methods of Crump (1984), which are sometimes referred to as the linearized multistage (LMS) model. Although q_1 is sometimes zero, q_1^* is always positive, and the upper limit on the extra risk of cancer (above the spontaneous incidence) associated with a small dose, d, has approximately the linear form $q_1^* d$.

Krewski et al. (1991) proposed a "model-free" estimate (MFX) of low-dose carcinogenic potency based on a series of secant approximations to the slope of the dose-response curve between points in the low-dose region and controls. Because their derivations are similar, MFX and q_1^* generally give similar estimates of low-dose carcinogenic potency (Krewski et al., 1991).

CORRELATIONS

Several authors have reported a high correlation between the TD_{50} (or carcinogenic potency) and the HDT within various selected subsets of data in the CPDB (Bernstein et al., 1985; Crouch et al., 1987; Rieth and Starr, 1989a,b). In most cases, the HDT was also the EMTD; that implies a high correlation between high-dose carcinogenic potency and potency in causing other chronic toxic effects. Zeise et al. (1984, 1985, 1986) and Metzger et al. (1989) have reported high correlations between the TD_{50} and the LD_{50} (i.e., between carcinogenic potency and acute toxicity).

The committee decided to review those correlations in an effort to investigate the relationship between toxicity and carcinogenic potency in MTD bioassays. Krewski et al. (Appendix F) performed the review and extended the reported correlations in several ways on the basis of data from their subset of 191 chemicals in the CPDB. They included in their analysis all studies in the CPDB that met the following criteria:

- Rodents were used.
- Chemicals were given orally.
- Results specified organ or tumor type, not total tumor-bearing animals.

- Exposure to the test chemical did not notably reduce survival of the test animals in comparison with unexposed controls.
- The dose-response trend was significant at $p < 0.01$.
- Authors stated that results were positive for carcinogenicity.
- The study included at least two doses and controls.

In their analysis, Krewski et al. omitted data at the highest dose if the dose-response curve turned downward and used the smallest TD_{50} if data were available from multiple sites or experiments. Krewski et al. estimated the TD_{50} with three models of the dose-response relationship: single-stage, multistage, and Weibull models. The correlation coefficients between estimates of the TD_{50} and HDT were 0.924, 0.952, and 0.821, respectively. Krewski et al. attributed the differences in correlation coefficients to the fact that the multistage model provides for upward curvature of the dose-response relationship, whereas the Weibull model provides for both upward and downward curvature and so is likely to permit a greater range of TD_{50} values. Krewski et al. also calculated correlations between the HDT and estimates of low-dose carcinogenic potency; they reported a correlation coefficient of -0.941 between the HDT and q_1^* and a correlation coefficient of -0.960 between the HDT and the estimate of low-dose potency based on the MFX. Finally, Krewski et al. explored how estimates of low-dose and high-dose carcinogenic potency could be predicted from the HDT on the basis of the observed correlations. Using the method of Gaylor (1989), they showed that a preliminary estimate of the upper-bound dose corresponding to the 95% upper confidence limit for an increased cancer risk of 1 x 10^{-6} based on the LMS model could be made in the absence of a standard bioassay by dividing the MTD by 380,000.

The main issue that has arisen in interpreting the observed correlations, both in previous publications (Bernstein et al., 1985; Crouch et al., 1987; and Rieth and Starr, 1989a,b) and in the MTD workshop discussions, is the extent to which the correlations are tautologous, that is, determined by features of the experimental designs and by the ways in which the experimental data are selected and analyzed, rather than by the underlying biologic mechanisms.

Figure 2-1 shows the TD_{50}s calculated by applying the one-stage dose-response model to data in the CPDB (Krewski et al., Appendix F) plotted against MTDs on a log-log scale. (The horizontal axis is the

highest dose tested, which is assumed to correspond to an estimate of the
MTD.)   As Figure 2-1 indicates, the data on the 191 chemicals are
tightly grouped about the best-fitting linear regression line; none of the
TD_{50}s appears to differ from that predicted by the regression line by
more than a factor of about 10.

Bernstein et al. (1985) point out that this relationship can be explained

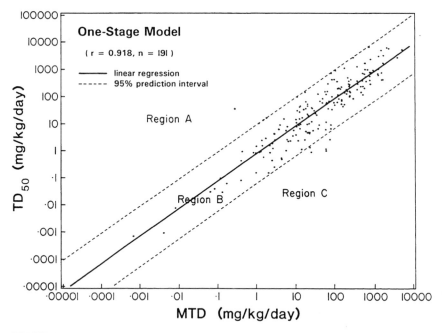

FIGURE 2-1   Association between carcinogenic potency and maximum
tolerated dose. Relationship between the highest dose tested for 191
chemicals that tested positive for carcinogenicity in chronic rodent bioassays and their TD_{50}, an inverse measure of carcinogenic potency defined
dose rate that halves likelihood of remaining tumor-free. Region A corresponds to chemicals of low carcinogenic potency (high TD_{50}) relative to
MTD; Region C corresponds to chemicals of high carcinogenic potency
(low TD_{50}) relative to MTD. Most chemicals fall into Region B, their
toxicities and carcinogenic potencies are correlated.

on the basis that, given the MTD, the TD_{50} is constrained to lie between two bounds. The lower bound is determined by the number of animals tested at the MTD, the spontaneous tumor rate, and the level of statistical significance required to label a chemical as a carcinogen. For example, if there is a 10% spontaneous rate and 50 animals are tested at the MTD, at least 10 animals must respond at the MTD if an effect is to be declared statistically significant at the 5% probability level. That minimum corresponds to a maximum TD_{50} value of 5.9 times the MTD. The upper bound is determined by the fact that it is very rare for 100% of the animals tested at the MTD to get tumors. If 49 of 50 animals get cancer at the MTD, compared with five of 50 control animals, the TD_{50} is estimated as 0.18 MTD. Thus, within those two bounds, the TD_{50} differs from the MTD only by, at most, a factor of about 6—i.e., 5.9 or 5.6 (1/0.18). Bernstein et al. (1985) showed that similar bounds apply to more general experimental designs involving two or three dosed groups. Assuming that the TD_{50} is uniformly distributed within those limits, Krewski et al. (Appendix F) showed that the theoretical correlation between log MTD and log TD_{50} would be 0.965, which is very close to the 0.918 obtained by Krewski et al. from the data shown in Figure 2-1. Thus, given the bounds established by Bernstein et al. (1985), a high correlation between MTD and TD_{50} is inevitable.

The committee's discussion of the possible interpretations and implications of those findings centered around the three regions shown in Figure 2-1. Region B (the region between the two broken lines) corresponds to the region determined by the bounds introduced by Bernstein et al. (1985); Region A corresponds to chemicals of low carcinogenic potency (high TD_{50}) relative to their MTD; and Region C corresponds to chemicals of high carcinogenic potency (low TD_{50}) relative to their MTD. The salient feature of this plot is that most of the chemicals in the analysis fall in Region B, whereas Regions A and C are virtually empty. Because the existence of a correlation is implied by the absence of chemicals in Regions A and C, the committee undertook to understand more fully why Regions A and C are nearly empty.

Krewski et al. (Appendix F) restricted their analysis to all chemicals in the CPDB that were clearly carcinogenic. Chemicals that were not identified as carcinogenic in any animal bioassay might have included both true noncarcinogens and chemicals with a carcinogenic potency that was too low to cause statistically significant increases in tumors in animals exposed at the MTD for a lifetime. Although the latter chemicals cannot be positioned with precision in Figure 2-1, it is clear from the definition of the three regions that they are Region A chemicals; if they could be positioned in Region A, the observed correlations would probably be reduced. Thus, the committee concludes that the correlation between the MTD and the TD_{50} might apply not to all chemicals, but only to those with carcinogenic potency high enough to cause statistically significant increases in tumors in animals exposed at the MTD for a lifetime. Therefore, the observed correlation is partially tautologous, to the extent that it might result partially from our inability to position Region A chemicals in graphs like Figure 2-1.

However, the absence of chemicals from Region C is not obviously tautologous. If a chemical tested in a standard long-term bioassay is a true Region C chemical, it should be identifiable as such. If a chemical caused cancer in all animals tested at the MTD, it still might not cause cancer in all animals exposed at lower doses or for shorter periods, in which case it could be positioned in Figure 2-1. If a chemical caused tumors in all the dosed animals in a bioassay, it could be identified as belonging in Region C, although it would not be possible to position it at a specific location in Region C. However, it is possible that chemicals belonging to Region C have been systematically excluded from the CPDB or, if present in the CPDB, excluded from the analyses conducted by Krewski et al. Although the inclusion criteria used by Krewski et al. (e.g., inclusion only of studies that used the oral route) resulted in the inclusion of only a fraction of the chemicals in the CPDB, it does not appear that there would be a bias toward excluding chemicals with low TD_{50}s relative to their HDTs. But the inclusion criteria for the CPDB could have resulted in the omission of some Region C chemicals. For instance, the exclusion of studies that lasted for less than half the normal lifespan of the animals could have resulted in excluding studies that were terminated when tumors were detected very early. And some chemicals might have been identified as potent carcinogens long ago and consequently not tested in a bioassay that satisfied standards used by Gold et al. (1984) as criteria for inclusion; e.g., some early bioassays might not have used control groups or might not have reported results in terms of the numbers of animals with tumors.

To investigate further the extent to which Region C carcinogens exist, the committee compiled a list of 18 chemicals (Table 2-1) that it judged as potential Region C carcinogens and conducted a detailed study of

TABLE 2-1   Chemicals Considered as Potential Region C Carcinogens

2-acetylaminofluorene[a]	Dimethyl sulfate[c]
Acrylonitrile[a]	Ethylene dibromide[a]
Benzidine[a]	Ethylene oxide[a]
Benzo[a]pyrene[b]	Ethylnitrosourea[a]
1,3-Butadiene[b]	Methyl bromide
Carbon tetrachloride[a]	4,4'-Methylene-Bis(2-chloroaniline)
C.I. Direct Black 38[a]	(MOCA)[a]
C.I. Direct Blue 6[a]	Plutonium[c]
C.I. Direct Brown 95[a]	Vinyl chloride[c]
Dibenz[a,h]anthracene[c]	

[a]Reported in CPDB.
[b]Additional data obtained.
[c]Not analyzed.

them. The detailed results of the study are included as Appendix G to this report. Twelve of the 18 chemicals were already represented in the CPDB. Suitable quantitative data were found on benzo[a]pyrene and 1,3-butadiene that permitted estimation of TD_{50}s for them, and an additional ingestion study of vinyl chloride was identified that was not included in the CPDB; data on those three chemicals were provided to Krewski et al., who calculated TD_{50}s with the same procedure as used in the workshop paper. TD_{50}s could not be estimated for dibenz[a,h]-anthracene, dimethyl sulfate, and methyl bromide, because no suitable quantitative data were located. Plutonium was not included in the analysis, because the dose measure used for it was not commensurate with that used for chemical carcinogens. Thus, of the 18 chemicals identified by the committee as potential Region C carcinogens, 14 were investigated. Five of these—the three benzidine dyes (C.I. direct black 38, C.I. direct blue 6, and C.I. direct brown 95), carbon tetrachloride, and 1,3-butadiene—belong in Region C (Figure 2-2). However, none of the TD_{50}s for these five chemicals was more than a factor of 10 or so smaller than the bound that divides Region C from Region B. Thus, the committee did not uncover any chemicals that appear to be positioned substantially far out into Region C.

The methods used by Krewski et al. (Appendix F) to estimate TD_{50}s

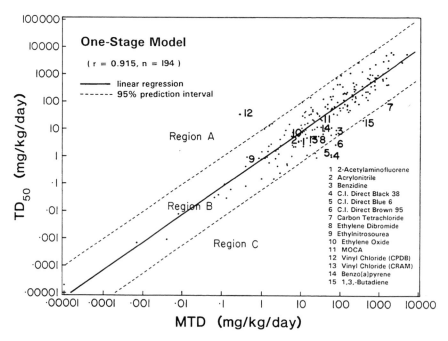

FIGURE 2   Association between carcinogenic potency and maximum tolerated dose. This graph is essentially the same as Figure 1 with 14 potential Region C chemicals explicitly identified. Included are data on three chemicals that were examined by the committee but were not present in the CPDB.

are similar to those used in the CPDB. As in the CPDB, the TD_{50}s in Figures 2-1 and 2-2 are based on an essentially linear one-stage model applied to the crude proportions of animals that developed tumors during the course of the study. For studies that lasted less than the standard lifespan of the test species, the TD_{50} is adjusted to a standard rodent lifetime by multiplying it by a correction factor, f^2, where f is the ratio of the length of the experiment to the normal lifespan of the test species. That correction is based on the assumption that, if experimental animals had lived longer, a greater percentage of them would have developed tumors as a result of exposure to the test agent, or agent-related tumors would have been discovered at lower doses.

The studies of the three benzidine dyes involved rats and lasted only 3 months, so a correction factor of $(3/24)^2 = 0.016$ was applied. Although the criteria for inclusion of rodent experiments in the CPDB stipulate that they lasted for a minimum of 12 months, all NTP studies are included in the CPDB, regardless of duration. The three experiments with the benzidine dyes were actually subchronic toxicity studies in which neoplastic lesions were observed. For ethylene dibromide, most of the rats died of tumors within 6 months, and a correction factor of $(6/24)^2 = 0.06$ was used. The carbon tetrachloride study lasted 84 weeks and involved a correction factor of $(84/104)^2 = 0.65$. Because the bioassays of each of the chemicals lasted less than the lifespan, the exact positioning of the five chemicals in Region C is uncertain (although more nearly certain for carbon tetrachloride).

The correction used in the CPDB is ad hoc and does not have a strong experimental or theoretical basis. The Environmental Protection Agency typically uses a correction factor of f^3 (Anderson et al., 1983), which would make estimated TD_{50}s even smaller. Portier et al. (1980) found that a factor of f^3 was consistent with many bioassays conducted by the NTP. Doll (1971) observed that a factor of f^2 to f^6 is needed to describe the age-incidence curves for human cancer. It is not clear what is an appropriate correction factor, so TD_{50}s estimated on the basis of studies lasting less than the standard lifespan of the test species are more uncertain than those estimated on the basis of whole-lifespan studies. The effect of this adjustment factor on the correlation between the TD_{50} and the MTD warrants investigation, although ultimate resolution of this uncertainty would require retesting of these chemicals at lower doses for a full lifespan.

In summary, the committee's informal study suggested that documented Region C carcinogens are rare. The best candidates for Region C carcinogens are five chemicals that induce cancer in rodents after short exposures. No bioassay lasting the full lifespan of the test species has been conducted for any of the five chemicals. Consequently, estimates of the TD_{50}s for the chemicals are uncertain, and their designation as Region C carcinogens is also uncertain.

In interpreting those results, we must keep two facts in mind. First, the TD_{50}s are uncertain for chemicals that have not been tested in lifetime bioassays; it is theoretically possible that these chemicals could be positioned substantially further out into Region C than is indicated by Figure 2-2. Second, the analyses assume that the HDT in each study is a reasonable approximation of the MTD; this might not be the case for all studies in the CPDB, particularly some of the older studies.

On the basis of the evidence discussed above, the committee concludes that the chemicals tested to date in lifetime bioassays have been found generally not to have TD_{50}s that are far less than their MTDs. The Bernstein lower bound of 0.18 MTD for the TD_{50} will not be universally applicable, however, because it is based on the use of an essentially linear one-hit model and the assumption that not all of the exposed animals will develop tumors. The lower bound also might not apply to TD_{50}s adjusted for intercurrent mortality with the method described by Sawyer et al. (1984), as is done in the CPDB whenever individual animal survival times are available. Krewski et al. (Appendix F) provide bounds on the TD_{50} relative to the MTD by using a Weibull dose-response model, which allows for curvature in the dose-response relationship. Although the bounds are wider, the correlation between the TD_{50} and MTD remains high, regardless of the degree of curvature in the dose-response curve (Kodell et al., 1990).

The committee further concludes that comparisons between TD_{50}s and MTDs indicate a relationship between measures of general toxicity (e.g., the MTD) and carcinogenic potency that can be expressed as follows: Animal carcinogens generally have a carcinogenic potency sufficient to cause just-detectable increases in cancer in standard bioassays at doses near the MTD.

The correlations considered here demonstrate that the carcinogenic potencies of materials found to be carcinogens are inversely related to MTDs; that is, if the MTD is high, the carcinogenic potency tends to be low. However, the MTD itself does not predict the likelihood that a material will be a carcinogen. The underlying cause of the observed relationship is not clear. General toxicity and cancer induction have a number of steps in common. A material must be absorbed, possibly metabolized, and transported to the site of action. Those common elements might be partially responsible for the observed relationship.

In addition, cell toxicity might result in increased cell division, which in turn could result in the permanent incorporation of spontaneous DNA damage that will eventually lead to cancer. The observed relationship between toxicity (as quantified by the MTD) and carcinogenicity (as quantified by the TD_{50}) is consistent with cell toxicity and the resulting cell proliferation's mediating of the carcinogenicity observed in some animal bioassays. However, the committee recognizes that other reasons for the observed relationship are possible. The committee suggests that experiments in which cell proliferation and carcinogenic or precarcinogenic responses are measured directly and compared will allow more definitive evaluation of relationships among toxicity, cell proliferation, and carcinogenicity.

RELATIONSHIP BETWEEN TOXICITY AND CARCINOGENICITY OBSERVED AT MTD

The practice of assessing risk associated with human exposures to chemicals on the basis of data from studies conducted in laboratory animals rests on a number of assumptions. Among them are the assumptions that the agents will produce qualitatively similar effects in animals and humans and that the relative potency in animals approximates the relative potency in humans. In general, assumptions about the relationships between animal and human data have proved fairly reliable. For instance, the application of toxicity, pharmacokinetic, and metabolic data derived from animal studies to human medicine has contributed to reducing the human risk associated with therapeutic agents.

The practice of classifying chemical substances as either carcinogenic or noncarcinogenic on the basis of animal tests conducted at the MTD involves a further assumption—that carcinogenesis is a specific response to exposure to specific chemical structures (agent specificity), rather than a nonspecific response of animals to induction of chronic toxicity. That assumption is necessary because chronic administration at the MTD often produces adverse effects in the tested animal populations. In fact, if no adverse effects have been observed in a chronic bioassay, the bioassay could be classified as inadequate, on the grounds that the MTD was not achieved and that the test had insufficient sensitivity to detect the carcinogenicity of the material tested. However, some researchers have argued that the observation of increased frequencies of tumors in animals receiving the MTD might not always be a chemical-specific phenomenon, but might be a secondary response to the induction of chronic toxicity. That is, perhaps chronic toxicity itself or some other high-dose phenomenon is capable of inducing cancer.

It has been suggested in particular that carcinogenic responses to exposures at high doses are in many cases either totally or partially caused indirectly by mitogenesis (Ames and Gold, 1990). The idea is that high doses (at or near the MTD) cause toxic responses, which can cause cell proliferation (mitogenesis). A dividing cell is at greater risk of mutating than a quiescent cell, so mitogenesis is indirectly mutagenic and consequently associated with an increased likelihood of carcinogenesis. That mechanism might be totally responsible for a carcinogenic response, as hypothesized for sodium saccharin (Cohen and Ellwein, 1990a), or might accentuate the carcinogenicity of genotoxic compounds, as hypothesized for 2-acetylaminofluorene (2-AAF) (Cohen and Ellwein, 1990b). In the former case, a threshold was hypothesized for saccharin on the basis of chemical evidence that silicate crystals responsible for cell proliferation in rats do not form at lower doses. In the latter case, a synergistic effect between genotoxicity and cell proliferation was hypothesized for 2-AAF at high doses in the bladders of female mice, but only a genotoxic effect at lower doses at which cell proliferation was not expected to occur. That observation suggests a dose-response relationship for bladder cancer that is nonlinear at high doses but linear at lower doses where cell proliferation is absent. 2-AAF does not induce cell proliferation in all target organ systems, however; the dose-response relationship for liver cancer in mice appears to be linear throughout the entire dose range.

The relationship between toxicity (including mitogenesis) and carcinogenesis has been studied recently. A direct relationship between toxicity and carcinogenesis has been suggested for a number of nongenotoxic chemicals, such as saccharin (noted above), the antioxidant butylated hydroxyanisole (BHA), di-(2-ethylhexyl) phthalate (DEHP), and polychlorinated biphenyls (PCBs). Chronic rodent bioassays of those chemicals have revealed tumor induction at doses that also are associated with toxicity and the presence of nonneoplastic proliferative lesions. For example, in two-generation studies in adults and weanling rats (Anderson et al., 1988; Williams, 1988), saccharin administered at 5% of the diet induces bladder tumors, cytotoxicity, and regenerative hyperplasia, increasing the labeling index (a measure of cell proliferation) of the urothelium by a factor of 2-10. BHA induces forestomach carcinomas in rats and hamsters when administered as 2% of the diet; severe hyperplasia and cytotoxicity, as evidenced by erosion and ulceration, are also seen (Ito et al., 1991). DEHP is a peroxisome proliferator that has been shown to induce liver tumors, foci of hepatocellular alteration (previously described as neoplastic nodules), and an initial burst of mitosis in rats and mice when given at 3,000-12,000 ppm in the diet (Kluwe et al., 1982; Mitchell et al., 1985). Some PCB mixtures induce focal necrosis, fatty degeneration, and hyperplastic nodules in the livers of rats and mice at concentrations that also induce hepatic adenomas and carcinomas (Kimbrough and Linder, 1974; Kimbrough et al., 1975).

In addition to nongenotoxic carcinogens, genotoxic carcinogens induce toxicity, and consequent cell proliferation at higher, toxic doses might play a role in increasing tumor rates to beyond what would be expected from genotoxicity alone. For example, 2-AAF administration is associated linearly with DNANTP adduct formation in the mouse bladder; however, the tumor rate in that organ is consistent with the effects of an increased rate of cell proliferation at high doses that acts in combination with 2-AAF's genotoxicity to produce tumors (Cohen and Ellwein, 1990). A similar interactive effect between cell proliferation and tumorigenesis has been observed for benzo[a]pyrene applied to mouse skin (Albert et al., 1991). Epidemiologic evidence also supports an association between some kinds of chronic toxicity and cancer incidence, such as hepatitis and liver cancer, schistosomiasis and bladder cancer, tuberculosis and lung cancer, asbestosis and mesothelioma, and tropical ulcers and skin cancer (Preston-Martin et al., 1991). Explanations other than cellular proliferation (such as inflammation) are also possible.

Thus, there is evidence from various sources to support an association between toxicity and carcinogenesis. Several people have recently attempted to analyze the assumption that the phenomena are causally related. Hoel et al. (1988) and Tennant et al. (1991) have evaluated the relationships between mutagenicity, carcinogenicity, and toxicity in laboratory rodents with the NTP data base of chronic and, in some cases, subchronic bioassays performed on a total of 130 chemicals. In those bioassays, 50 rats and mice of each sex received the MTD, MTD/2, or MTD/4 for 2 years. Matched control groups were also used. Use of the data base provided an opportunity to compare the toxic properties of chemicals that were not carcinogenic with those of chemicals that

were, under similar experimental conditions. Results included subchronic toxicity, neoplastic and chronic toxic effects observed after a 2-
year exposure, chemical structure, and mutagenicity in salmonellae.
Toxicity was defined by the investigators as "any deleterious change in
the tissues of animals exposed to chemicals that was discerned by histo-
pathology"; most, but not all, toxic lesions were found to be associated
with a proliferative response. Qualitative structural descriptors of toxicity were used to evaluate the relationships between regenerative or hyperplastic responses and cancer or the absence of cancer; rates of induced
mitogenesis or increased rates of cell proliferation were not measured
quantitatively and would have remained undetected in the absence of any
structural change.

Table 2-2 shows the chemicals from the Tennant et al. (1991) study
whose subchronic and chronic administration induced toxicity at the
same site; some of the chemicals were carcinogens and some were not,
but none was carcinogenic at the site of obvious toxicity. In contrast,
Table 2-3 lists the chemicals that were carcinogenic at sites where both
subchronic toxicity and chronic toxicity were present; about 40% of
these were mutagenic. For both the concordant and discordant chemicals, most of the toxic lesions observed were proliferative, although the
presence of proliferative lesions clearly is not predictive of carcinogenesis. Results of the Tennant et al. (1991) analysis and the Hoel et al.
(1988) analysis indicate that some sites of toxicity of both carcinogens
and noncarcinogens were associated with neoplasia and many were not.
Some chemicals induced tumors at sites where toxicity was not in evidence, and some induced toxicity in some organs without inducing carcinogenesis. However, the majority of both mutagenic and nonmutagenic
carcinogens induced tumors that were associated with chronic toxicity,
although many of the same chemicals caused chronic toxicity at other
sites that was not associated with carcinogenesis. Tennant et al. (1991)
conclude that, although their results do not dissociate toxicity from the
neoplastic process, they "illustrate the high degree of complexity of
neoplastic processes and imply that there may be multiple mechanisms
of carcinogenesis associated with even potent mutagens. They also
provide a clear demonstration that chronic exposure of rodents to chemicals that exhibit toxic effects does not necessarily result in carcinogenic
effects. Further, even when chronic exposures resulted in overt tissue-
specific toxicity, neoplasia did not necessarily develop." A temporary

TABLE 2-2 Chemicals That Induced Both Subchronic and Chronic Toxicity but Not Carcinogenicity at Same Sites[a]

Chemical	Rats		Mice	
	Site	Lesion	Site	Lesion
Noncarcinogens				
2,4-Dichlorophenol	None	--	Liver	Syncytial alteration
Dimethoxane	Forestomach	Hyperplasia	Forestomach	Hyperplasia
Hydrochlorothiazide	Kidney	Nephropathy, mineralization	None	--
α-Methyldopa sesquihydrate	None	--	Kidney	Nephropathy, karyomegaly
Carcinogens				
p-Chloroaniline HCl	Bone marrow, liver	Hyperplasia, hemosiderin	Kidney	Hemosiderin
Nitrofurantoin	Testes	Degeneration	Testes	Degeneration
Tribromoethane	Liver	Inflammation, vacuolization	Liver	Inflammation, vacuolization
Malonaldehyde, Na salt	Glandular stomach; bone marrow	Ulcer, inflammation; hyperplasia	Pancreas	Atrophy
Furosemide	None	--	Kidney	Nephropathy

[a]From Tennant et al. (1991).

TABLE 2-3 Chemicals That Induced Toxicity and Carcinogenicity at Same Site[a]

Chemical	Mutagen	Rats		Mice	
		Site	Lesion	Site	Lesion
Glycidol	+	Forestomach	Dysplasia, carcinoma	Forestomach	Hyperplasia, carcinoma
p-Chloroaniline HCl	+	Spleen; adrenal	Fibrosis, metaplasia, sarcoma; hyperplasia, pheochromocytoma	Liver	Hemosiderosis, hepatocellular tumors
N,N-Dimethylaniline	+	Spleen	Fibrosis, metaplasia, sarcoma	Forestomach	Hyperplasia, squamous cell tumors
Nitrofurantoin	+	Kidney	Nephropathy, tubular cell adenoma and carcinoma	Ovary	Atrophy, ovarian tumors
4-Vinyl-1-cyclohexene diepoxide	+	Skin	Hyperplasia, basal and squamous cell carcinoma	Skin, ovary	Hyperplasia, basal and squamous cell tumors; atrophy, ovarian tumor
N-Methylolacrylamide	-	None	--	Ovary	Atrophy, granulosa cell tumors
Benzofuran	-	Kidney	Nephropathy, tubular cell adenocarcinoma	Liver; forestomach; lung	Syncytial changes, liver tumors; hyperplasia, carcinoma; hyperplasia, carcinoma

TABLE 2-3 Continued.

Chemical	Mutagen	Rats		Mice	
		Site	Lesion	Site	Lesion
Ochratoxin A	-	Kidney	Degeneration, hyperplasia, tubular cell tumors	Not determined	--
Hexachloroethane	-	Kidney	Nephropathy, hyperplasia, tubular cell adenocarcinoma	Not determined	--
d-Limonene	-	Kidney	Mineralization, nephropathy, hyperplasia, tubular cell adenocarcinoma	None	--
Hydroquinone	-	Kidney	Nephropathy, tubular cell adenoma	Liver	Syncitial cell alteration, liver tumors
Phenylbutazone	-	Kidney	Papillary necrosis, nephropathy, transitional-cell carcinoma	Liver	Degeneration, hypertrophy, necrosis, liver tumors

[a] From Tennant et al. (1991).

toxic condition's effect on carcinogenesis might not be detected with data from chronic or even subchronic bioassays; for example, it is possible that a chemical very early in the course of a bioassay induces toxicity that enhances its carcinogenic response, but that, because of an adaptive cellular response, no chronic proliferative lesions other than tumors develop. Nonetheless, the observations that have been reported after study of the NTP database support the existence of mechanisms of carcinogenesis more complex than simple mutation or induced cell proliferation; these mechanisms are yet to be identified.

Several other reports support the conclusion of an equivocal relationship between toxicity-induced proliferation and carcinogenesis discussed above. Wada et al. (1990) showed that p-methoxyphenol administered after initiation of rat forestomach tumors with N-methyl-N'-nitro-N-nitrosoguanidine caused epithelial damage and hyperplasia in a dose-dependent manner in the forestomach epithelium, but was not associated with any increase in tumors. In an investigation of the role of renal tubular cell hyperplasia in tumor promotion with barbital sodium (BBNa) after initiation with streptozotocin (STZ) in rats, STZ was found to reduce BBNa-induced nephropathy and cell proliferation without reducing renal-tumor incidence (Konishi et al., 1990). The authors of the study note, however, that initiated cells might have a very different ability from noninitiated cells to respond to the mitogenic influences of a tumor promoter and that the reduction in overall DNA synthesis that was seen might be unrelated to the increased proliferation of preneoplastic or neoplastic cells. Ward et al. (1990) reached a similar conclusion in a study of the relationship between renal or hepatocellular hyperplasia and tumor promotion with di-(2-ethylhexyl)phthalate in mice initiated with N-nitrosoethylurea.

The observation that toxicity and carcinogenicity are not always detected simultaneously make it problematic to account for increased rates of cell proliferation that are associated with carcinogenesis when one performs risk assessments of either genotoxic or nongenotoxic chemicals. The greater-than-quadratic nature of many dose-response curves for mutagens tested at and below their MTDs and the observation of toxicity and proliferative lesions in the target organs of most mutagenic carcinogens suggest that mechanisms in addition to mutation are operative and, in particular, that enhanced cell proliferation is likely to be occurring and affecting the response. In addition, most nonmutagens also induce toxicity and nonneoplastic proliferative lesions at doses that also are associated with neoplasia. Any information on the dose-response nature of these effects, especially cell proliferation, should be included in assessments of risk where possible, although, as Konishi et al. (1990) and Ward et al. (1990) emphasize, the target cells for proliferative activity associated with carcinogenesis might not be the total parenchymal tissue; identification of the affected target cells, such as stem cells, could be necessary. These problems are addressed in the second part of this report, *Issues in Risk Assessment: Two-Stage Models of Carcinogenesis.*

In summary, the committee evaluated the likelihood that observed correlations between cancer potencies and other measures of toxicity of chemicals tested at the MTD are tautologous and result from bioassay design or the statistical methods used for analysis or have a biologic basis. The committee performed its evaluation by determining the correlation between estimates of the TD_{50} and the HDT of clearly carcinogenic chemicals found in the CPDB. A strong correlation between those quantitites was observed, with no chemicals classifiable as having either high toxicity and low potency or low toxicity and high potency. The committee concluded that the correlation is partly tautologous because it applies only to chemicals with cancer potencies high enough to be detected in an MTD bioassay. However, the relationship is not entirely tautologous, possibly because the phenomena of toxicity and carcinogenesis have several similarities. The dichotomy is reflected in the conflicting results of Tennant et al. (1991), who reported that an association between some measures of toxicity and positive carcinogenicity results in the same target organ in some, but not all, NTP bioassays. It is not yet possible to draw further conclusions about the relationship between toxicity and cancer potency.

3

Advantages and Disadvantages of Bioassays that Use the MTD

QUALITATIVE INFORMATION

The current animal bioassay was designed as a qualitative screen for carcinogenicity and noncarcinogenicity. It typically does not yield information about the carcinogenicity of materials at doses much below the EMTD (e.g., lower than MTD/2). Current bioassay designs usually include a control group, a group exposed at the MTD, and one to two additional doses, the lowest of which is MTD/10 to MTD/2. Thus, testing is generally conducted in a relatively narrow range of doses below the MTD. Effects observed at these doses might or might not be relevant to human exposure at environmental concentrations.

There are a number of advantages to including the MTD in long-term rodent bioassays. The MTD is the dose most likely to induce tumors; as a result, its use provides information about tumor type and about which organs are sensitive to the carcinogenic effects of a test substance. This information can provide a basis for designing followup studies for characterizing the biologic mechanisms through which cancer is produced. Knowledge of target organs can also assist epidemiologists in designing studies among human populations exposed to a chemical, although species can differ in the tissue that responds to a chemical.

When bioassays are conducted in more than one animal species, use of the MTD provides a consistent basis for interspecies comparisons. By evaluating the differences in response in different species, strains, and sexes of animals, we improve our ability to extrapolate the data from the animal tests to humans. For the few chemicals on which data exist, potency estimates calculated from animal data have been reported to be highly correlated with similar estimates made from human data; this increases confidence that animal data can be used to predict results in humans (Allen et al., 1988).

Another advantage of the MTD is that tests benefiting from the greater sensitivity of the MTD are more likely to give positive or negative results that can be a starting point for structure-activity correlation analyses. In addition, use of the MTD provides information on a number of end points of toxicity in addition to cancer.

A disadvantage of animal bioassays as they are currently performed is that they generally are not designed specifically to provide information on biochemical and physiologic mechanisms operating during the production of tumors. However, the MTD bioassay might yield clues about mechanisms, which can aid in designing mechanistic studies. For example, preliminary, short-term testing is conducted before the bioassay, primarily to determine how much of a chemical can be administered to animals without decreasing their lifespan through causes other than cancer (i.e., to estimate the MTD). Consequently, mild, non-life-threatening toxicity is common in groups of animals receiving the EMTD. It can lead to changes in food consumption, recurrent cytotoxicity in specific organs, hormonal imbalance, or combinations of these and other effects. Those effects have been associated with both increases and decreases in tumor incidence in laboratory animals (Reitz et al., 1980, 1990; Turnbull et al., 1985; Roe, 1988) and could be used to provide the first clues to an agent's biologic mechanisms. Those mechanisms are discussed below.

If a chemical alters physiologic processes the alterations can influence delivery of the chemical or its metabolites to a target site or its clearance from a target site. An example is the effect of large lung burdens of particles on clearance of particles inhaled later (Lee et al., 1985). When animals are exposed to high concentrations of insoluble particles, their lungs rapidly accumulate particle burdens large enough to overwhelm the lungs' normal clearance mechanisms; this increases the rate of accumulation of particles inhaled later. The phenomenon has been reported for diesel-soot particles (Mauderly et al., 1987) and titanium dioxide (Lee et al., 1985). In such a situation, lung tumors might be induced in laboratory animals even by particles that are nontumorigenic at lung burdens acquired in more relevant human exposures. It is not clear whether this secondary process exhibits a predictable dose-response relationship.

The magnitude of exposure to xenobiotic compounds is known to affect the pathways by which they are metabolized. Metabolic enzymes can be characterized by their affinity for substrates and their capacity for metabolizing them. At low doses, high-affinity low-capacity enzymes can be expected to play the major role in metabolism of a chemical; at high doses, low-affinity high-capacity enzymes will be major contributors. If the metabolic pathway of a chemical is a low-capacity pathway and produces the carcinogenic metabolite, as is true for the metabolism of benzene (Sabourin et al., 1990) and vinyl chloride (Maltoni et al., 1981), exposures of animals to high doses of the chemical, particularly in bolus doses, can lead to an underestimation of the potential for tumor production at lower doses.

If the high-affinity low-capacity metabolic enzymes play a protective role in an organism, and the low-affinity high-capacity enzymes produce reactive, potentially toxic intermediates, then administration of high doses might cause a shift in metabolism to the more toxic pathway. For example, administration of high doses of methylene chloride to mice causes a disproportionate increase in metabolism by the glutathione transferase pathway, and Andersen et al. (1987) suggested that production of reactive metabolites from this pathway was responsible for induction of liver tumors in mice. Similarly, Reitz et al. (1984a,b) noted that, when conjugation of the male rat bladder tumorigen *o*-phenylphenol (Hiraga and Fujii, 1981, 1984) was saturated by administration of high doses, a shift in metabolic pathways was associated with production of more reactive metabolites. In these cases, overestimation of the potential for tumor production at lower doses would be expected.

Several chemicals are known to induce hormonal imbalances, which might in turn induce tumors by secondary mechanisms. Many of those compounds alter Phase I or Phase II metabolizing enzymes (Sipes and Gandolfi, 1991) or both, and some—such as diethylstilbestrol, methyltestosterone, zearalenones, and retinoids—mimic endogenous hormones (Nebert, 1991). In addition, a small number of nonmutagenic, carcinogenic chlorinated polycyclic hydrocarbons, such as 2,3,7,8-tetrachlorodibenzo-*p*-dioxin (TCDD) and polychlorinated biphenyls (PCBs), exhibit hormone-like activity through receptor-driven mechanisms. One of the primary cellular actions of TCDD is through a cytosolic nuclear protein designated as the Ah (aryl*h*ydrocarbon) receptor, which is analogous in some ways to steroid hormone receptors in both structure and function (Poland and Knutson, 1982; Umbreit and Gallo, 1988). The binding of the ligand to the receptor activates the receptor, which translocates to the nucleus, binds to a response element on the DNA, and serves as a trans-acting growth regulatory factor (Whitlock, 1986; Landers and Bunce, 1991). Receptor theory and recent evidence supporting hormone and TCDD receptor theory (Stephenson, 1956; Jordan and Murphy, 1990; DeVito et al., 1991) support the argument that the process might require complex occupancy of multiple receptors and nuclear binding sites and that the process is compound-specific and saturable (Safe, 1986). As with hormone action, there is some reason to believe that responses to low doses of TCDD-like molecules might not be sufficient to trigger physiologic responses, but high-dose exposures to receptor ligands could lead to aberrant growth, cytotoxicity, and increased risk of cancer. Work now in progress, however, suggests that receptor binding might not by itself explain all the dose-response effects observed at low doses (G. Lucier, NIEHS, pers. comm., 1992).

Some chemicals (thioureas and sulfonamides) are thought to produce thyroid neoplasia secondary to hormonal imbalance in rodents through a variety of mechanisms, including altered thyroid hormone synthesis or hormonal metabolism (Paynter et al., 1988; Hill et al., 1989). Other chemicals that produce thyroid tumors are hepatic microsomal enzyme inducers in rodents at high doses and alter thyroid function by increasing the hepatic disposition of thyroid hormone (Oppenheimer et al., 1968; Cavalieri and Pitt-Rivers, 1981; Hill et al., 1989). Decreased serum thyroid hormone concentrations could result in a compensatory increase in pituitary TSH, which in turn might exert a tumor-promoting effect (Hiasa et al., 1982) or an increase in thyroid neoplasia (McClain, 1989). Because small amounts of thyroxine will block the tumor-promoting effect of a microsomal enzyme inducer, such as phenobarbital, this effect is presumed to be secondary to hormonal imbalance, as opposed to a direct tumor-promoting or direct carcinogenic effect in the thyroid (Mc-Clain et al., 1988, 1989; McClain, 1989).

Reserpine has been reported to produce adrenal medullary tumors in male rats, mammary tumors in female mice, and seminal vesicle tumors in male mice (DHEW, 1979). All three tumor types might be secondary to the neurogenic effects of reserpine. Reserpine increases cell proliferation in the adrenal medulla (Tischler et al., 1988, 1989); that this can be prevented by unilateral denervation of the adrenal gland indicates that the cell proliferation is probably a neurogenically mediated reflexive response to catecholamine depletion (Tischler et al., 1991). Reserpine and other neuroleptic agents increase serum prolactin via inhibition of dopaminergic neurotransmission in the hypothalamus. Thus, the mammary tumors might be secondary to increased serum prolactin.

The examples noted above indicate that failure to account for biologic mechanisms of action of many chemicals that elicited tumors when they were tested in bioassays at their MTDs could lead to errors in qualitative and quantitative predictions about human carcinogenesis.

The specificity and sensitivity of animal bioassays are also important considerations for evaluating the predictability of bioassay results for humans. *Sensitivity* refers to the ability of bioassays to detect true human carcinogens, and *specificity* refers to their ability to avoid mistaking substances that do not cause cancer in humans for carcinogens. The sensitivity of bioassays, as well as can be determined at present, is very high. All the known human carcinogens adequately tested thus far (about 39) have been carcinogenic in one or more animal species (Huff and Rall, 1992), although target organs are often inconsistent among species. The specificity of animal bioassays, however, cannot be evaluated now because information on human noncarcinogens is insufficient to make comparisons. The default assumption has been that evidence supporting or refuting carcinogenic activity in animal studies should be considered applicable to humans until better information is obtained.

The most likely explanation for the small numbers of carcinogens and noncarcinogens identified in humans is the relative insensitivity of clinical and epidemiologic methods and their great difficulty in demonstrating causality (Karstadt et al., 1981). It is common for epidemiologists to propose associations between exposures to environmental substances and later cancer formation, and for experimentalists to demonstrate biologic plausibility and causality of the associations through animal studies.

An alternative explanation for the finding of many more rodent carcinogens than human carcinogens is that animal bioassays conducted at the MTD are too sensitive or of higher sensitivity than human epidemiologic studies of exposure at lower doses. One way of examining the question of sensitivity is to consider the responses of animals to fractions of the MTD. Among the difficulties in doing so are that the EMTD might be an overestimate or underestimate of the true MTD, that the fractions of the highest dose that are tested are still high doses, and that, as the dose is reduced, the statistical power to detect an effect is also reduced. Hoel et al. (1988) reported that, among a group of 52 animal carcinogens tested at multiple doses, 34 (65%) showed statistically significant effects at lower doses. All but three of the remaining 18 substances also increased tumor incidences in a lower-dose group, compared with incidences in controls, but the increases were not statistically significant.

Another possible explanation is that the high proportion of carcinogens found in animal studies reflects bias in the selection of substances for testing. Substances can be selected for a variety of reasons, such as widespread human exposure, commercial use, or prior suspicion of carcinogenicity. In one study in which prior suspicion of carcinogenicity was evaluated as an important selection criterion, Griesemer (NIEHS, pers. comm., 1991) found that, of 255 substances tested because they were suspected of carcinogenicity, 169 (66%) were carcinogenic in animals; of 127 substances tested for other reasons, 26 (20%) were carcinogenic in animals.

One consequence of such selection bias is relative confidence in our ability to identify noncarcinogens. If increased cancer incidences are not detected in animals exposed to a substance at the MTD, one might conclude that the substance is noncarcinogenic for the species-sex-strain combination being tested (within the limits of sensitivity of the test), or that the carcinogenic potential of the substance is too low for carcinogenicity to be detected (under the conditions of the test). Failure to observe statistically significant increases in tumors in a standard set (two species and both sexes) of bioassays performed at the MTD has become the operational definition of *noncarcinogen*. It is not possible to apply that operational definition in bioassays in which a dose substantially below the MTD was used as the highest dose tested. In such cases, use of a higher dose or the MTD might have revealed a carcinogenic effect. Of course, chemicals that satisfy the operational definition of *noncarcinogen* might be shown to have carcinogenic potential if tested in larger numbers of animals or in additional strains or species. But the definition has proved useful in categorizing chemicals for regulatory purposes.

The idea that substances that are carcinogenic at very high doses might not be carcinogenic at lower doses requires the assumption that high doses of chemicals perturb the body or its defensive mechanisms in a qualitatively different manner from lower doses of the same substances. The MTD is deliberately designed to be close to the lower border of toxicity, so it is logical to consider whether some aspect of toxicity promotes carcinogenicity. Few studies have been conducted to address that issue directly, i.e., by testing proposed mechanistic hypotheses in long-term animal studies. In two analyses of NTP studies, Hoel et al. (1988) and Tennant et al. (1991) reported only partial correlation between the sites and types of toxic effects measured in conventional toxicologic studies and the development or lack of development of cancer. It is possible that a less conspicuous component of toxic responses, such as changes in mutation rates, or toxic responses measured much earlier in the bioassay, such as induced cell proliferation, will provide supporting evidence that toxicity provokes cancer, but much more research will be needed.

QUANTITATIVE INFORMATION

In addition to indicating whether a chemical is a carcinogen in rodents at high doses, a test performed at the MTD yields information about the carcinogenic potency of the chemical in rodents. Potency is a function of both the dose and the magnitude of the observed response. One chemical is judged to have a higher potency than another if the percentage of animals that develops tumors at a given dose is higher than for the other chemical. Current procedures typically involve testing about 50 animals per sex per species per dose. If the background incidence of tumors were 10% (5/50) in controls, the minimum statistically significant response at the MTD would be 20% (10/50). The highest response would, of course, be 100% (50/50). Clearly, a 100% response would indicate a higher potency for the test chemical than a 20% response. Knowledge about the quantitative response obtained in the bioassay enables scientists to make estimates of relative carcinogenic potency and adds information beyond the simple identification of a substance as a likely carcinogen or noncarcinogen.

Current bioassays that use the MTD and one or two lower doses provide some limited information about the shape of the high-dose portion of the dose-response curve. The shape of the curve at high doses might or might not have relevance to low-dose exposures, however, depending on the reliability of the qualitative assumptions that have been made. The current standard regulatory practice of estimating "plausible upper bounds" on risk often applies the linearized multistage model (Crump, 1984) to bioassay data obtained with the MTD. That procedure is based on the assumption of a linear relationship between tumor response and dose at low doses. Because of the linearity assumption, estimates of low-dose risk obtained with this procedure are often not very sensitive to the observed dose-response shape in the experiment. It is important to note that the linear relationship cannot be verified directly, nor can it be verified that the estimate so computed provides a true upper bound on risk at very low doses. Nevertheless, the linearized multistage procedure has been widely used by regulatory agencies and is thought to provide an objective basis for decisions concerning regulation of chemicals found, in high-dose bioassays, to increase tumor frequencies in animals. In particular, the procedure allows a crude rank ordering of animal carcinogens from most potent to least potent, which might then provide a basis for priorities in regulation and pollution prevention.

One of the reasons that MTD bioassays provide little information on the shape of the dose-response curve is that it uses a small number of doses. The same number of groups and numbers of animals per group tested at lower doses would probably yield even less information about the shape of the dose-response curve or about carcinogenicity, however, because testing at lower doses decreases the likelihood of response in a small experiment or the likelihood that a response will be distinguishable from background. If the response cannot be distinguished from background, no information useful for defining the shape of the curve is obtained. Although most earlier NTP and NCI studies exposed animals only at the MTD and MTD/2, more recent studies have also exposed animals at MTD/4. Among 38 positive responses in sex- and species-specific studies, 23 (61%) would have been positive if MTD/4 had been the largest dose. Seven of the 23 would have detected all site-specific responses, and the other 16 would have detected some, but not all, site-specific effects (R. Griesemer, NIEHS, pers. comm., 1991); even the use of low experimental doses in standard bioassays can provide useful dose-response information (i.e., result in responses that are distinguishable from background). Information on the shape of the dose-response curve below the range of the chronic bioassay is more likely to be obtained from experiments targeted at elucidating biologic mechanisms of action and characterizing the dose-response characteristics of the critical events, such as DNA-adduct formation or induced cell proliferation.

There are cases, of course, where extrapolating from high to low doses can be irrelevant. Some human exposures levels can be directly compared with dosing at the MTD; and in some cases, the dose rate might be comparable, especially with high-dose occupational exposures (Gold et al., 1987b) and many pharmaceutical exposures. In these cases, the human dose rates are very similar to those used in rodent bioassays, and the focus of supplementary testing would be to elucidate biologic mechanisms to determine the human relevance of bioassay results, not the validity of low-dose extrapolation.

Another problem with the use of the MTD is that, although it is now included in most carcinogenicity bioassays, criteria for selecting the EMTD and evaluating the selection vary among laboratories. In bioassays conducted by the NTP, the highest dose tested is the EMTD estimated from a 90-day study, and sufficient data are presented to determine whether the MTD was achieved. However, other published bioassay results might not state the rationale for dose selection and might insufficiently report toxicity data and other data needed to determine whether the MTD was achieved.

The usefulness of reports of bioassay results for regulators and risk managers could be increased by including several pieces of information: a clearly stated rationale for dose selection and a summary of the toxicity information important in evaluating the dose selection, especially whether the animals could have tolerated a higher dose and whether the high dose used elicited toxicity. Such considerations should be included in a risk assessment of a substance. The potential for reduced statistical power of studies in which the MTD was not achieved, compared with studies in which the MTD was achieved, should be noted and accounted for, or at least acknowledged in a risk assessment.

4

Options Considered

The committee considered several options relative to the use of the MTD as the highest dose for use in carcinogenicity screening studies. These options were initially proposed by the participants in the MTD workshop organized by the committee in consultation with the federal liaison group. The first option would retain the status quo, with the possible addition of lower doses in addition to the MTD. The second option would use a high dose that is an arbitrary fraction of the EMTD. The third option would redefine the MTD, basing it on studies of the dose dependence of physiologic effects expected to alter carcinogenic response. The fourth option would use MTD testing as part of an overall testing strategy that separates carcinogens from noncarcinogens and provides information useful for determining human relevance; this could take one of two forms—a two-track system that comprises full testing and limited testing and a system of sequential studies. These options are presented below and are followed by discussions of their advantages and disadvantages.

OPTION 1

Continue carcinogenicity screening studies with the MTD as the highest dose according to current practice (with the inclusion of lower doses as well).

This option has the advantage that the MTD bioassay is the only currently standardized method in the United States for identifying carcinogens. Continuing the bioassay as it is currently used allows comparisons with the results of similar studies conducted in the past. The test is designed to achieve high sensitivity. Compounds found to be negative in a standard set (two species and both sexes) of bioassays conducted at the MTD can be designated as noncarcinogens with a relatively high degree of confidence.

The negative aspects of the test are that its results indicate only whether a chemical is a rodent carcinogen under the conditions of the assay. The current screening system is oriented toward identifying potential carcinogens. Without additional studies, the bioassay does not indicate how predictive the results of the rodent bioassay are for humans, and it provides little information with which to estimate responses at low doses—which might be of particular concern to humans. For materials for which the evidence of carcinogenicity is weak (i.e., related to only one sex-species group or to the highest dose) and that are economically important, further studies to elucidate the metabolism or mechanisms of action, particularly regarding effects on cell proliferation, are in order.

OPTION 2

Redesign the bioassay so that the highest dose is some small fraction of the EMTD.

When human exposures were reasonably anticipated to be within a factor of 10 or 100 of the animal MTD or when the test substance was judged to have a high potential for direct interaction with DNA (as judged by results of short-term tests for genotoxicity), the MTD would continue to be used as the highest dose.

For materials that did not meet those criteria, however, the MTD would not be routinely used as the highest dose in chronic bioassays. Instead, a fraction of the EMTD (e.g., MTD/3) would serve as the highest dose for bioassays. Additional doses would be spaced geometrically below the high (MTD/3) dose at half-log intervals (i.e., the second dose would be MTD/10, the third dose would be MTD/33, etc.).

If the numbers of animals in a test were not substantially increased, agents that were formerly found to be carcinogenic only at the MTD would not be identified as positive in this procedure. Thus, this procedure would focus attention on agents that had high carcinogenic potency relative to their subacute toxicity (Region C carcinogens in Figure 1).

An advantage of this procedure is a flexibility that reflects differences between potentially hazardous substances themselves or the expected exposures of humans to them. Most such chemicals will never be tested in chronic animal bioassays; current evidence suggests that, if they were, about half would be identified as carcinogens and almost all, by definition, would be in Region B. It would clearly pose a dilemma for regulators to be faced with decisions on a multitude of chemicals that would be identified as carcinogens under today's regulatory standards. However, substances with low toxicity but high carcinogenic potency (Region C carcinogens) might well present an unusually high cancer risk to human populations but not produce toxicity in the bioassay that serves to warn people of a hazard; that is apparently what happened with vinyl chloride workers (Fox and Collier, 1977; IARC, 1979).

A disadvantage of this option is that it would decrease the sensitivity of the assay, thus reducing its usefulness as a means of hazard identification. This disadvantage can be compensated for by increasing the numbers of animals in test groups; however, the expense of increasing the number of animals to a point necessary to retain the same power would probably be prohibitive. Furthermore, the choice of a lower dose, such as MTD/3, as the highest dose is arbitrary. Although implementation of this option would identify primarily Region C carcinogens, there is little evidence that Region C carcinogens contribute a predominant portion of chemically induced cancer risk to human populations. Future uses of a chemical cannot always be anticipated. If the HDT were based on current uses, a future use that entailed high human exposures could necessitate a new bioassay.

OPTION 3

Base the HDT on preliminary studies that determined the dose dependence of physiologic effects induced by the chemical and the dose dependence of its metabolism.

In this option, a comprehensive series of tests would be conducted before the bioassay were initiated. The tests would be designed to provide information about mechanisms of toxicity, as well as the dose-response curve for such toxicity. Microscopic examination of tissues would continue to be a part of studies, as would clinical chemistry (e.g., serum enzyme measurements and urinalysis.) However, the studies would be expanded to include measurement of physiologic and biochemical effects (e.g., alterations in hormone status) and quantitative measurements of cell proliferation. In addition, extensive pharmacokinetic studies (quantitative measurements of uptake, distribution, metabolism, and elimination) would be carried out.

An expert panel would be convened to evaluate preliminary data before doses for the bioassay were selected. The panel would select the HDT and lower doses on the basis of evaluation of the preliminary studies. The objective would be to design a study that could be expected to yield results that would be useful for human risk assessment and not simply to administer as much chemical as possible without causing early mortality from causes other than cancer. This approach constitutes a change in emphasis of the bioassay. Studies that use the MTD as currently defined are designed to maximize the sensitivity of the bioassay (i.e., to prevent false-negative results). The objective of the new approach would be to increase the selectivity of the bioassay (i.e., to decrease the number of false-positive results).

In some cases, adoption of this option would not change the selection of the MTD as the HDT. For example, if human populations were reasonably expected to encounter high concentrations of the test substance, the MTD would continue to be used. In many cases, however, the HDT would be lower than the MTD as currently defined, and the spacing of doses could be much wider than commonly adopted by programs such as the National Toxicology Program (NTP).

An advantage of this modification is that the mechanisms underlying any observed carcinogenic response would be more likely to be qualitatively and quantitatively similar to those operating at lower doses than mechanisms underlying responses observed at the MTD.

A disadvantage of this modification might be that doses that caused a physiologic change in one organ might not cause physiologic changes in other organs. Without knowledge of the target sites for carcinogenicity, it would be unclear whether a physiologic change that is being avoided by testing at lower doses has any relevance to those other target sites. For example, if an HDT were selected to be below a dose that caused a physiologic change in the liver but not the lung, one could miss a carcinogenic effect on the lung at a dose that does not alter lung physiology.

OPTION 4

Use MTD testing as part of an overall testing strategy.

The animal bioassay that uses the MTD is one part of a complete program for identifying human carcinogens. It generally is conducted after some indication that a substance merits examination—e.g., information that a chemical has a structural similarity to a known carcinogen, results of a test for mutagenicity, or a suggestion that there will be extensive human exposure to the substance. It is then used as a screening technique to separate carcinogens from noncarcinogens it can be followed by tests to determine mode of action, pharmacokinetics, and applicability of results to the human experience. The workshop participants and the committee discussed two ways to use the MTD test in a complete program. They are described below.

Option 4A

Use a two-track system that comprises full testing and limited testing.

In this option, chemicals would be placed into two tracks for testing. A small number of selected chemicals would be subjected to rigorous testing (the full-testing track). All the remaining chemicals would be subjected to less rigorous testing (the limited-testing track). The option is based on three premises:

- A large amount of additional information might be needed to assist in understanding the importance of positive results at the MTD.
- It might not be feasible to collect the additional information for all the chemicals whose regulation is appropriate.

• Without accompanying information on mechanisms or results at low doses, animal bioassay results alone (i.e., without parallel data on mechanisms and dose-response relationships) do not add greatly to our ability to make regulatory decisions because of the uncertainty about the human implications of positive results in animal bioassays.

Chemicals could be chosen for full testing on the basis of expected human exposures, importance in commerce, structural similarity to a known carcinogen, or results of mutagenicity tests—i.e., in much the same way that chemicals are currently chosen for testing. The method of Gaylor (1989), basing a preliminary estimate of the carcinogenic potency of a chemical on its MTD, could also be used to select chemicals for full or limited testing. As described earlier, given the relationship between the carcinogenic potency of chemicals and their MTDs, Gaylor pointed out that a preliminary estimate of the dose corresponding to a carcinogenic risk of one in a million human lifetimes could be found by dividing the MTD by 380,000. If human exposures were unlikely to be greater than the quotient, the chemical would not be assigned to full testing on the grounds that, even if it were a carcinogen, human risk would be unlikely to be greater than one in a million per lifetime. (A different divisor could be selected if a level of safety different from one in a million per lifetime were required.)

When a class of structurally similar chemicals that are thought likely to have similar mechanisms of action is being considered, it might be a good use of resources to test fully only a small number (perhaps only one) of representative chemicals in the class and to evaluate the others in the class for relative potency on the basis of data from short-term, less-expensive studies.

Chemicals chosen for full testing would be tested in a standard bioassay that used the MTD and an array of doses below the MTD. If a chemical were positive, additional testing would be performed as needed to clarify the dose-response relationship and to improve the predictive value of the positive findings for humans.

Chemicals chosen for limited testing would be considered for regulation without testing in a long-term cancer bioassay. Regulatory decisions for these chemicals would be based on more limited data, such as estimates of the MTD from short-term studies, mutagenicity information, and other data that can be gathered much more quickly and cheaply than results from a lifetime bioassay.

Such a strategy could best be implemented by a program, such as the NTP, that has general responsibility for assaying a large number of chemicals. Special-interest groups concerned with chemicals assigned to the limited-testing track would have the option of conducting more rigorous tests.

An advantage of this approach is that it would provide a strong data base for evaluating carcinogenic potency for the few chemicals that undergo full testing. It also has the potential to permit more effective use of testing resources.

A disadvantage is that chemicals in the limited-testing track might be evaluated inadequately, and that might result in overregulation or underregulation of individual chemicals. Furthermore, criteria for assigning a chemical to a track might not be reliable. For example, mutagenicity does not always correlate with carcinogenicity, and some structurally similar chemicals might not have similar mechanisms of action.

Option 4B

Perform sequential studies.

In this option, pharmacokinetic studies would be included as part of the early short-term toxicity studies that are used to determine the EMTD. Relatively simple pharmacokinetic studies could determine the approximate dose that exceeds the ability of an animal to absorb and metabolize the test chemical. The usefulness of such data in the design of a long-term study can be illustrated by the pharmacokinetics of inhaled methyl bromide (Medinsky et al., 1985). Doubling the exposure concentration from 5,700 to 10,400 nmol/L in a 6-hour exposure did not increase the internal dose received by rats. The higher concentration caused a decrease in minute volume and in the percentage absorbed, the animals did not receive any more of the test compound than at the lower concentration. Obviously, there would be no point in designing a long-term study at a concentration that the test animals could not absorb. Similar studies could determine the dose at which the animals' capacity to metabolize the internal dose is overwhelmed, as indicated by increased excretion of the parent chemical. With such information, long-term studies could be designed that include at least one dose that does not exceed the metabolic capacity of the animal.

After the short-term toxicity studies, which would be used to find both the EMTD and the pharmacokinetic characteristics of the test chemical, the long-term bioassay would be conducted with the MTD and lower doses, one of which did not exceed the capacity of the animals to absorb and metabolize the chemical. Additional animals might be required for the latter dose. If the results of the long-term bioassay were negative, no further studies would be conducted. If the results were positive, studies to determine the relevance of the animal responses to human risk would be conducted. Such studies would be aimed at determining the shape of the dose-response curve for events that can lead to cancer. Additional work might include more detailed pharmacokinetic studies, such as studies of the effect of dose on the kinetics of specific metabolic pathways and the identification of metabolites; studies of the effects of the test chemical on primary physiologic control systems, such as the endocrine, renal, and cardiovascular systems; studies of the effects on growth-regulating systems in target organs and cells (such as perturbation of oncogene products, alteration of protein kinases, activation of cytokines, and alteration of hormones); studies of mechanisms of mutagenesis; and studies of the induction or reduction of detoxifying enzyme systems. Epidemiologic evidence of carcinogenicity would also be important in determining the relevance of the animal response to human risk. The results of those studies, in conjunction with bioassay data, should provide the framework needed to understand the events that lead to cancer, improve predictability across species (particularly from rodents to humans), and provide a biologic basis for low-dose extrapolation.

An advantage of this approach is that it systematically contributes information needed for risk assessment.

A disadvantage is that it will take longer to complete the evaluation of a chemical, and it will be possible to test only a few chemicals in such a thorough program.

5

Conclusions and Recommendations

Several decades ago, there was no standard bioassay for detecting chemical carcinogens. The current MTD bioassay in rodents closed that gap and became the standard assay in the United States. It is neither perfect nor unalterable, and by itself it is insufficient to produce data from which accurate human-health risk assessments can be made. Nonetheless, the MTD bioassay does provide some useful information for hazard identification and risk assessment.

The committee began its deliberations by examining the question of agent specificity—i.e., whether carcinogenicity seen in animal bioassays at the MTD is a specific response to a chemical, rather than a response to general toxicity. Few studies have been conducted to address the issue directly by testing proposed mechanistic hypotheses in long-term animal studies. Recent studies have reported an inconsistent relationship between the sites and types of chronic toxic effects measured in conventional toxicicity studies and the development or lack of development of cancer. Although toxicity and carcinogenicity often are not detected simultaneously, nonspecific toxicity-induced carcinogenesis can occur, and increased rates of cell proliferation associated with carcinogenesis (when they occur) should be considered in risk assessments of genotoxic and nongenotoxic chemicals.

The committee concludes that the relationship between measures of general toxicity (e.g., the MTD) and carcinogenic potency can be expressed as follows: Potency of animal carcinogens is such that the increase in tumor incidence that is needed to show statistical significance in standard bioassays occurs generally at doses near the MTD, but use of the MTD itself does not predict whether a material will elicit a carcinogenic response in a standard animal bioassay. The basis of the relationship is not clear. General toxicity and cancer induction have a number of steps in common. A material must be absorbed, possibly metabolized, and transported to a site of action. The sharing of those elements might be partially responsible for the observed relationship. In addition, the relationship is consistent with chemical-induced toxicity's being a mediator of carcinogenicity in some instances. However, the committee recognizes that other reasons for the relationship are also possible.

Because of the relationship between TD_{50}s and MTDs of chemicals, the committee concludes that a preliminary upper bound on the potential carcinogenic potency of an untested chemical can be estimated from knowledge of its MTD. Such estimates can prove useful in setting priorities for carcinogenicity testing and in making preliminary upper-bound risk estimates of cancer risk whenever carcinogenicity bioassay results are not available. If such an upper-bound estimate predicts a low human risk, a chemical could be assigned a low priority for carcinogenicity testing.

The committee noted that, although testing at the MTD is currently included in most carcinogenicity bioassays, specific criteria for selecting the EMTD and evaluating the selection vary with the group conducting the study.

The committee recommends that, to facilitate interpretation, reports of bioassay results should include a clearly stated rationale for dose selection and a summary of the toxicity information important for evaluating the dose selection.

The committee concludes that the MTD bioassay as currently conducted in rodents is most useful as a qualitative screen to determine whether a chemical has the potential to induce cancer. It also provides information on the carcinogenic potential of a substance at high doses and some information about the dose-response curve. It does not provide (nor was it intended to provide) all the information useful for low-dose human risk assessment. In most cases, additional information is likely to be needed to determine the extent to which the induction of cancer in rodents adequately predicts the human response and how the results of the relatively high-dose assay can best be used to make inferences about the expected effects at low doses. Some of the required information might be obtainable from study of tissues from animals subjected to long-term bioassay or from ancillary studies incorporated in the rodent bioassay. Future bioassays should be designed to reveal the overall toxic responses induced by the test chemical and not just the carcinogenic response. But in general, other information needs will require other studies.

The types of additional information that are needed will depend on the chemical under study, but several general subjects merit consideration. Studies should be conducted that can help to determine whether the mechanisms or metabolic processes involved in the production of cancer in rodents are relevant to humans. In addition, toxicokinetic studies are important to determine the effect of dose on the absorption and metabolic fate of the chemical. Physiologic responses induced by the chemical and the effect of dose on that response must be considered. Furthermore, chemically induced cell proliferation of target cells might play an important role in the induction of some tumors, and cell proliferation should be measured when that is appropriate. It is difficult to conceive how similar information could be gathered in humans.

The committee could not reach consensus on how additional information on mechanisms of carcinogenicity and other responses should be used in conjunction with the MTD bioassay, however. A majority of the committee believes that the assay identifies substances that do or do not increase the incidence of cancer under the conditions of the assay and provides an operational definition of *animal noncarcinogens* The assay also identifies target organs, demonstrates tumor types associated with exposure, provides a consistent basis for interspecies comparisons, and can serve as a guide in designing followup studies. The majority also believes that there are as yet no validated mechanisms of carcinogenicity that support lowering the MTD and that failing to use the MTD for carcinogen screening will reduce the sensitivity of the bioassay and diminish the opportunity to compare results among chemicals and species. As a result, the majority recommends implementation of Option 4B, described earlier in the report. This option is summarized as follows.

The MTD should continue to be one of the doses used in carcinogenicity bioassays. Other doses, ranging downward

from MTD/2 possibly to MTD/10 or less, should also be used. The capacity of the test animal to absorb and metabolize the test chemical should be taken into account in selection of doses below the MTD.

If a standard set of rodent bioassays that each include the MTD are negative, generally no additional tests related to carcinogenicity are required.

If a bioassay conducted with the MTD is positive, additional studies should be performed to reduce uncertainties in predicting human responses to the test material and to assist in performing quantitative risk assessments.

These additional studies should address mechanisms of cancer induction, toxicokinetics and metabolism of the material, and and physiologic responses induced by the material; they could also include validation of MTD bioassay results with epidemiologic studies.

Some committee members disagreed with those two recommendations and believe that a greater modification of the process for selecting doses to be used in carcinogenicity bioassays is required. The modification would emphasize specificity over sensitivity and would require that bioassay doses be selected after careful evaluation of subchronic studies conducted before the chronic bioassay, as discussed earlier in this report as Option 3. The minority recommends the following:

Bioassay doses should be selected by a panel of experts on the basis of careful evaluation of studies conducted before the bioassay is initiated. Information gathered before the bioassay is conducted would include information about mechanisms of toxicity in test animals and an elucidation of the dose-response relationship for such toxicity. The HDT should be chosen as the highest dose that can be expected to yield results relevant to humans, not the highest dose that can be administered to animals without causing early mortality from causes other than cancer.

The committee is aware that regulation of a chemical can take place after any stage in data collection. Public-health considerations can lead to cautious behavior, sometimes expressed as formal regulation early in the testing history of the chemical. Such regulation can be expected to use conservative (cautious) assumptions, which should be eased or otherwise modified as the accumulation of data makes estimation of the dose-response relationship more precise or improves knowledge of the pharmacokinetics of the material and its mode of action.

References

Agency for Toxic Substances and Disease Registry (ATSDR). 1990. Toxicological Profile for Polycyclic Aromatic Hydrocarbons. TP-90-20. U.S. Department of Health and Human Services, Washington, D.C.

Albert, R.E., M.L. Miller, T.E. Cody, W. Barkley, and R. Shukla. 1991. Cell kinetics and benzo[a]pyrene-DNA adducts in mouse skin tumorigenesis. Pp. 115–122 in Chemically-Induced Cell Proliferation: Implications for Risk Assessment, B.E. Butterworth, T.J. Slaga, W. Farland, and M. McClain, eds. New York: Wiley-Liss.

Allen B., K.S. Crump, and A.M. Shipp. 1988. Is it possible to predict the carcinogenic potency of a chemical in humans using animal data? Pp. 197–209 in Carcinogen Risk Assessment: New Directions in the Qualitative and Quantitative Aspects, R.W. Hart and F.D. Hoerger, eds. Banbury Report #31. Cold Spring Harbor Laboratory, Cold Spring Harbor, N.Y.

Ames, B.N., and L.S. Gold. 1990. Too many rodent carcinogens: Mitogenesis increases mutagenesis. Science 249:970–971.

Anderson, R.L., and the Carcinogen Assessment Group of the U.S. Environmental Protection Agency. 1983. Quantitative approaches in use to assess cancer risk. Risk Anal. 3:277–295.

Andersen M.E., H.J. Clewell, III, M.L. Gargas, F.A. Smith, and R.H. Reitz. 1987. Physiologically based pharmacokinetics and the risk assessment process for methylene chloride. Toxicol. Appl. Pharmacol. 87:185–205.

Anderson, R.L., R.R. Lefever, and J.K. Maurer. 1988. Comparison of the responses of male rats to dietary sodium saccharin exposure initiated during nursing with responses to exposure initiated at weaning. Food Chem. Toxicol. 26:899–907.

Armitage, P., and R. Doll. 1961. Stochastic Models for Carcinogenesis. Pp. 19-38 in Proceedings of the 4th Berkeley Symposium on Mathematical Statistics and Probability, Vol. 4. Berkeley and Los Angeles: University of California Press.

Bernstein, L., L.S. Gold, B.N. Ames, M.C. Pike, and D.G. Hoel. 1985. Some tautologous aspects of the comparison of carcinogenic potency in rats and mice. Fundam. Appl. Toxicol. 5:79–86.

Cavalieri, R.R., and R. Pitt-Rivers. 1981. The effects of drugs on the distribution and metabolism of thyroid hormones. Pharmacol. Rev. 33:55–80.

Cohen, S.M., and L.B. Ellwein. 1990a. Cell proliferation in carcinogenesis. Science 249:1007–1011.

Cohen, S.M., and L.B. Ellwein. 1990b. Proliferative and genotoxic cellular effects in 2-acetylaminofluorene bladder and liver carcinogenesis: Biological modeling of the ED01 study. Toxicol. Appl. Pharmacol. 104:79–93.

Crouch, E.A.C., R. Wilson, and L. Zeise. 1987. Tautology or not tautology? J. Toxicol. Environ. Health 20:1–10.

Crump, K. 1984. An improved procedure for low-dose carcinogenic risk assessment from animal data. J. Environ. Pathol. Toxicol. Oncol. 5(4-5):339–348.

Danse, L.H., F.L. van Velsen, and L.A. van der Heijden. 1984. Methylbromide: Carcinogenic effects in the rat forestomach. Toxicol. Appl. Pharmacol. 72:262–271.

DeVito, M.J., T.H. Umbreit, T. Thomas, and M.A. Gallo. 1991. An Analogy Between the Actions of the Ah Receptor and the Estrogen Receptor for Use in the Biological Basis for Risk Assessment of Dioxin. Banbury Conference on the Biological Basis for Dioxin Risk. Cold Spring Harbor, N.Y.

DHEW (Department of Health Education and Welfare). 1979. Bioassay of Reserpine for Possible Carcinogenicity. DHEW Publication No. NIH 79-1749. Department of Health Education and Welfare, National Institute of Health, Bethesda, Maryland.

Diehl, J.H., R.A. Guilmette, B.A. Muggenberg, F.F. Hahn, and I.Y.

Chang. 1992. Influence of dose rate on survival time for ^{239}PuO$_2$-induced radiation pneumonitis or pulmonary fibrosis in dogs. Radiat. Res. 129:53-60.

Doll, R. 1971. The age distribution of cancer: Implications for models of carcinogenesis. J. R. Stat. Soc. A37:133-166.

EPA (U.S. Environmental Protection Agency). 1986. Health and Environmental Effects Profile for Methyl Bromide. Prepared by the Office of Health and Environmental Assessment and the Environmental Criteria and Assessment Office for the Office of Solid Waste and Emergency Response. Washington, D.C.

Feron, V., C.F. Hendriksen, A.J. Speek, H.P. Til, and B.J. Spit. 1981. Lifespan oral toxicity study of vinyl chloride in rats. Food Cosmet. Toxicol. 19:317-333.

Fox, A., and P. Collier. 1977. Mortality experience of workers exposed to vinyl chloride monomer in the manufacture of polyvinyl chloride in Great Britain. Brit. J. Ind. Med. 344:1-10.

Gaylor, D.W. 1989. Preliminary estimates of the virtually safe dose for tumors obtained from the maximum tolerated dose. Regul. Toxicol. Pharmacol. 9:101-108.

Gold, L.S., C.B. Sawyer, R. Magaw, G.M. Backman, M. de Veciana, R. Levinson, N.K. Hooper, W.R. Havender, L. Bernstein, R. Peto, M.C. Pike, and B.N. Ames. 1984. A carcinogenic potency database of the standardized results of animal bioassays. Environ. Health Perspect. 58:9-21.

Gold, L.S., M. de Veciana, G.M. Backman, R. Magaw, P. Lopipero, M. Smith, M. Blumenthal, R. Levinson, L. Bernstein, and B.N. Ames. 1986a. Chronological supplement to the carcinogenic potency database: Standardized results of animal bioassays published through December 1982. Environ. Health Perspect. 67:161-200.

Gold, L.S., L. Bernstein, J. Kaldor, G. Backman, and D. Hoel. 1986b. An empirical comparison of methods used to estimate carcinogenic potency in long-term animal bioassays: Lifetable vs summary incidence data. Fundam. Appl. Toxicol. 6:263-269.

Gold, L.S., J.M. Ward, L. Bernstein, and B. Stern. 1986c. Association between carcinogenic potency and tumor pathology in rodent carcinogenesis bioassays. Fundam. Appl. Toxicol. 6:677-690.

Gold, L.S., T.H. Slone, G.M. Backman, R. Magaw, M. Da Costa, P. Lopipero, M. Blumenthal, and B.N. Ames. 1987a. Second chro-

nological supplement to the carcinogenic potency database: Standardized results of animal bioassays published through December 1984
and by the National Toxicology Program through May 1986. Environ. Health Perspect. 74:237-329.

Gold, L.S., G.M. Backman, M.K. Hooper, and R. Peto. 1987b.
Ranking the potential carcinogenic hazards to workers from exposures
to chemicals that are tumorigenic in rodents. Environ. Health Perspect. 76:211-219.

Gold, L.S., T.H. Slone, and L. Bernstein. 1989a. Summary of carcinogenic potency and positivity for 492 rodent carcinogens in the
carcinogenic potency database. Environ. Health Perspect. 79:259-
272.

Gold, L.S., L. Bernstein, R. Magaw, and T.H. Slone. 1989b. Interspecies extrapolation in carcinogenesis: Prediction between rats and
mice. Environ. Health Perspect. 81:211-219.

Gold, L.S., T.H. Slone, G.M. Backman, S. Eisenberg, M. Da Costa,
M. Wong, N.B. Manley, L. Rohrbach, and B.N. Ames. 1990.
Third chronological supplement to the carcinogenic potency database:
Standardized results of animal bioassays published through December
1986 and by the National Toxicology Program through June 1987.
Environ. Health Perspect. 84:215-286.

Haseman, J.K. 1985. Issues in carcinogenicity testing: Dose selection.
Fund. Appl. toxicol. 5:66-6678.

Henderson, B., and S. Preston-Martin. 1990. Increased cell proliferation as a cause of human cancer. In Proceedings of the Annual Meeting of the American Association of Cancer Research 31:513-515.

Hiasa, Y., Y. Kitahori, M. Ohshima, T. Fujita, T. Yuasa, N. Konishi,
and A. Miyashiro. 1982. Promoting effects of phenobarbital and
barbital on development of thyroid tumors in rats treated with N-
bis(2-hydroxypropyl) nitrosamine. Carcinogenesis 3:1187-1190.

Hill, R.N., L.S. Erdreich, O.E. Paynter, P.A. Roberts, S.L. Rosenthal,
and C.F. Wilkinson. 1989. Thyroid follicular cell carcinogenesis.
Fundam. Appl. Toxicol. 12:629-697.

Hiraga, K., and Fujii, T. 1981. Induction of tumors of the urinary
system in F344 rats by dietary administration of sodium o-phenyl-
phenate. Food Cosmet. Toxicol. 19:303-310.

Hiraga, K., and Fujii, T. 1984. Induction of tumors of the urinary
bladder in F344 rats by dietary administration of o-phenylphenol.

Food. Cosmet. Toxicol. 22:865-870.

Hoel, D.G., J.K. Haseman, M.D. Hogan, J. Huff, and E.E. Mc-Connell. 1988. The impact of toxicity on carcinogenicity studies: Implications for risk assessment. Carcinogenesis 9:2045-2052.

Huff, J., and D.P. Rall. 1992. Relevance to humans of carcinogenesis results from laboratory animals toxicology studies. Pp. 433-440 and 453-457 in Maxcy-Rosenau-Last Public Health and Preventive Medicine, 13th ed., J.M. Last and R.B. Wallace, eds. Norwalk, Ct.: Appleton & Lange.

IARC (International Agency for Research on Cancer). 1974. Some aromatic amines, hydrazine, and related substances, N-nitroso compounds and miscellaneous alkylating agents. Pp. 271-275 in IARC Monographs on the Evaluation of Carcinogenic Risk of Chemicals to Man, Vol. 4. Lyon, France: International Agency for Research on Cancer.

IARC (International Agency for Research on Cancer). 1979. Some monomers, plastic and synthetic elastomers, and acrolein. Pp. 377-438 in IARC Monographs on the Evaluation of the Carcinogenic Risk of Chemicals to Humans, Vol. 19. Lyon, France: International Agency for Research on Cancer.

IARC (International Agency for Research on Cancer). 1986a. Long-term and Short-term Assays for Carcinogens: A Critical Appraisal, R. Montesano, H. Bartsch, H. Vainio, J. Wilbourn, and H. Yamasaki, eds. IARC Science Publications #83. Lyon, France: International Agency for Research on Cancer. 575 pp.

IARC (International Agency for Research on Cancer). 1986b. Statistical Methods in Cancer Research, Vol. 3. The Design and Analysis of Long-term Animal Experiments, J.J. Gart, D. Krewski, P.N. Lee, R.E. Tarone, and J. Wahrendorf, eds. IARC Science Publications #79. Lyon, France: International Agency for Research on Cancer. 219 pp.

IARC (International Agency for Research on Cancer). 1991. Preamble. Pg. 21 in chlorinated drinking-water; chlorination by-products; some other halogenated compounds; cobalt and cobalt compounds. IARC Monographs on the Evaluation of Carcinogenic Risks to Humans, Vol. 52. Lyon, France: International Agency for Research on Cancer.

Ito, N., M. Hirose, and S. Takahashi. 1991. Cellular proliferation and stomach carcinogenesis induced by antioxidants. Pp. 43–52 in Chemically Induced Cell Proliferation: Implications for Risk Assessment, B.E. Butterworth, T.J. Slaga, W. Farland, and M. McClain, eds. New York: Wiley-Liss.

Jordan, V.C., and C.S. Murphy. 1990. Endocrine pharmacology of antiestrogens as antitumor agents. Endocr. Rev. 11:578–610.

Karstadt, M., R. Bobal, and I.J. Selikoff. 1981. A survey of availability of epidemiologic data on humans exposed to animal carcinogens. Pp. 223–245 in Quantification of Occupational Cancer, R. Peto and M. Schneiderman, eds. Banbury Report #9. Cold Spring Harbor Laboratory, Cold Spring Harbor, N.Y.

Kimbrough, R., and R.E. Linder. 1974. Induction of adenofibrosis and hepatomas of the liver in BALB/cJ mice by polychorinated biphenyls (Aroclor 1254). J. Natl. Cancer Inst. 53:547–552.

Kimbrough, R.D., R.A. Squire, R.E. Linder, J.D. Strandberg, R.J. Montalli, and V.W. Burse. 1975. Induction of liver tumors in Sherman strain female rats by polychlorinated biphenyl Aroclor 1260. J. Natl. Cancer Inst. 55:1453–1459.

Kluwe, K.M., J.K. Haseman, J.F. Douglas, and J.E. Huff. 1982. The carcinogenicity of dietary DEHP in Fischer 344 rats and B6C3F$_1$ mice. J. Toxicol. Environ. Health 10:797–815.

Kodell, R.L., D.W. Gaylor, and J.J. Chen. 1990. Carcinogenic potency correlations: Real or artifactual? J. Toxicol. Ind. Health. Submitted.

Konishi, N., B.A. Diwan, and J.W. Ward. 1990. Amelioration of sodium barbital-induced nephropathy and regenerative tubular hyperplasia after a single injection of streptozotocin does not abolish the renal tumor promoting effect of barbital sodium in male F344/NCr rats. Carcinogenesis 11:2149–2156.

Krewski, D., D.W. Gaylor, A.P. Soms, and M. Szyszkowicz. 1990 (Appendix B). Correlation Between Carcinogenic Potency and the Maximum Tolerated Dose: Implications for Risk Assessment. Paper prepared for the Committee on Risk Assessment Methodology, September 4, 1990, Washington, D.C.

Krewski, D., D. W. Gaylor, and M. Szyszkowicz. 1991. A model-free approach to low-dose extrapolation. Environ. Health Perspect. 90: 279–285.

Landers, J.P., and N.J. Bunce. 1991. The Ah receptor and the mechanism of dioxin toxicity. Biochem. J. 276:273–287.

Lee, K.P., H.J. Trochimowicz, and C.F. Reinhardt. 1985. Pulmonary response of rats exposed to titanium dioxide by inhalation for 2 years. Toxicol. Appl. Pharmacol. 79:179–192.

Maltoni, C., G. Lefemine, A. Ciliberti, G. Cotti, and D. Carretti. 1981. Carcinogenicity bioassays of vinyl chloride monomer: A model of risk assessment on an experimental basis. Environ. Health Perspect. 41:3–29.

Mantel, N., and W.R. Bryan. 1961. Safety testing of carcinogenic agents. J. Nat. Canc. Inst. 27:455–470.

Mauderly, J.L., R.K. Jones, W.C. Griffith, R.F. Henderson, and R.O. McClellan. 1987. Diesel exhaust is a pulmonary carcinogen in rats exposed chronically by inhalation. Fundam. Appl. Toxicol. 9:208–221.

McClain, R.M. 1989. The significance of hepatic microsomal enzyme induction and altered thyroid function in rats: Implications for thyroid gland neoplasia. Toxicol. Pathol. 17:294–306.

McClain, R.M., R.C. Posch, T. Bosakowski, and J.M. Armstrong. 1988. Studies on the mode of action for thyroid gland tumor promotion in rats by phenobarbital. Toxicol. Appl. Pharmacol. 94:254–265.

McClain, R.M., A.A. Levin, R.C. Posch, and J.C. Downing. 1989. The effect of phenobarbital on the metabolism and excretion of thyroxine in rats. Toxicol. Appl. Pharmacol. 99:216–228.

McConnell, E.E. 1989. The maximum tolerated dose: The debate. J. Amer. Coll. Toxicol. 8:1115–1120.

Medinsky, M.A., J.S. Dutcher, J.A. Bond, R.F. Henderson, M.B. Snipes, J.W. McWhinney, L.S. Birnbaum, and Y.S. Cheng. 1985. Uptake and excretion of carbon-14 methyl bromide as influenced by exposure concentration. Toxicol. Appl. Pharmacol. 78:215–225.

Melnick, R.L., J. Huff, B.J. Chou, and R.A. Miller. 1990. Carcinogenicity of 1,3-butadiene in C57BL/6 x C3HF$_1$ mice at low exposure concentrations. Cancer Res. 50:6592–6599.

Metzger, B., E. Crouch, and R. Wilson. 1989. On the relationship between carcinogenicity and acute toxicity. Risk Anal. 9:169–177.

Mitchell, A.M., J.C. Lhugenot, J.W. Bridges, and C.R. Elcombe. 1985. Identification of the proximate peroxisome proliferator(s) derived from DEHP. Toxicol. Appl. Pharmacol. 80:23–32.

Moolgavkar, S.H., and D.J. Venzon. 1979. Two-event models for carcinogenesis: Incidence curves for childhood and adult tumors. Mathemat. Biosci. 47:55-77.

Moolgavkar, S.H., and A.G. Knudson. 1981. Mutation and cancer: A model for human carcinogenesis. J. Natl. Cancer Inst. 66:1037-1052.

Neal, J., and R.H. Rigdon. 1967. Gastric tumors in mice fed benzo-(a)pyrene: A quantitative study. Tex. Rep. Biol. Med. 25:553-557.

Nebert, D.W. 1991. Proposed role of drug-metabolizing enzymes: Regulation of steady state levels of the ligands that affect growth, homeostasis, differentiation, and neuroendocrine functions. Mol. Endrocrin. 5:1203-1214.

Oppenheimer, J.H., G. Bernstein, and M.I. Surks. 1968. Increased thyroxine turnover and thyroidal function after stimulation of hepatocellular binding of thyroxine by phenobarbital. J. Clin. Invest. 47:1399-1406.

OTA (Office of Technology Assessment), U.S. Congress. 1987. Identifying and Regulating Carcinogens. OTA-BP-H-42. Washington, D.C.: U.S. Government Printing Office. November.

Paynter, O.E., G.J.Burin, R.B. Jaeger, and C.A. Gregorio. 1988. Goitrogens and thyroid follicular cell neoplasia: Evidence for a threshold process. Regul. Toxicol. Pharmacol. 8:102-119.

Peto, R., M.C. Pike, L. Bernstein, L. S. Gold, and B.N. Ames. 1984. The TD$_{50}$: A proposed general convention for the numerical description of the carcinogenic potency of chemicals in chronic-exposure animal experiments. Environ. Health Perspect. 58:1-8.

Poland, A., and J.C. Knutson. 1982. 2,3,7,8-tetrachlorodibenzo-p-dioxin and related halogenated aromatic hydrocarbons: Examination of the mechanism of toxicity. Annu. Rev. Pharmacol. Toxicol. 22:517-554.

Portier, C.J., S.C. Hedges, and D.G. Hoel. 1980. Age-specific models of mortality and tumor onset for historical control animals in the National Toxicology Program's carcinogenicity experiments. Cancer Res. 46:4372-4378.

Preston-Martin, S., M.C. Pike, R.K. Ross, and B.E. Henderson. 1991. Epidemiologic evidence for the increased cell proliferation model of carcinogenesis. Pp. 21-34 in Chemically Induced Cell Proliferation: Implications for Risk Assessment, B.E. Butterworth, T.J. Slaga, W.

Farland, and M. McClain, eds. New York: Wiley-Liss.

Reitz, R.H., P.G. Watanabe, M.J. McKenna, J.F. Quast, and P.J. Gehring. 1980. Effects of vinylidene chloride on DNA sysnthesis and DNA repair in the rat and mouse: A comparative study with dimethylnitrosamine. Toxicol. Appl. Pharmacol. 52:357–370.

Reitz, R.H., T.R. Fox, J.F. Quast, E.A. Hermann, and P.J. Watanabe. 1984a. Molecular mechanisms involved in the toxicity of orthophenylphenol and its sodium salt. Chem. Biol. Interact. 43:99–119.

Reitz, R.H., T.R. Fox, J.F. Quast, E.A. Hermann, H.D. Kirk, and P.J. Watanabe. 1984b. Biochemical factors involved in the effects of orthophenylphenol (OPP) and sodium orthophenylphenate (SOPP) on the urinary tract of male F344 rats. Toxicol. Appl. Pharmacol. 73: 345–349.

Reitz, R.H., A.L. Mendrala, R.A. Corley, J.F. Quast, M.L. Gargas, M.E. Andersen, D.A. Staats, and R. B. Conolly. 1990. Estimating the risk of liver cancer associated with human exposures to chloroform using physiologically-based pharmacokinetic modeling. Toxicol. Appl. Pharmacol. 105:443–459.

Reuzel, P.G., H.C. Dreef-van der Meulen, V.M. Hollanders, C.F. Kuper, V.J. Feron, and C.A. van der Heijden. 1991. Chronic inhalation toxicity and carcinogenicity study of methyl bromide in Wistar rats. Food Chem. Toxicol. 29:31–39.

Rieth, J.P., and T.B. Starr. 1989a. Chronic bioassays: Relevance to quantitative risk assessment of carcinogens. Regul. Toxicol. Pharmacol. 10:160–173.

Rieth, J.P., and T.B. Starr. 1989b. Experimental design constraints on carcinogenic potency estimates. J. Toxicol. Environ. Health 27:287–296.

Roe, F.J.C. 1988. How do hormones cause cancer? Pp. 259–272 in Theories of Carcinogenesis, O.H. Iversen, ed. New York: Hemisphere Publishing Corporation.

Sabourin, P.J., J.D. Sun, J.T. MacGregor, C.M. Wehr, L.S. Birnbaum, G.W. Lucier, and R.F. Henderson. 1990. Effect of repeated benzene inhalation exposures on benzene metabolism, binding to hemoglobin, and induction of micronuclei. Toxicol. Appl. Pharmacol. 103:452–462.

Safe, S. 1986. Comparative toxicology and mechanism of action of polychlorinated dibenzo-*p*-dioxins and dibenzofurans. Annu. Rev.

Pharmacol. Toxicol. 26:371–399.

Sawyer, C., R. Peto, L. Bernstein, and M.C. Pike. 1984. Calculation of carcinogenic potency from long-term animal carcinogenesis experiments. Biometrics 40:27-40.

Scott, B.R. 1988. Early and Continuing Effects of Combined Alpha and Beta Irradiation of the Lung: Phase II Report. Division of Regulatory Applications, Office of Nuclear Regulatory Research, U.S. Nuclear Regulatory Commission, Washington, D.C.

Scott, B.R., F.F. Hahn, M.B. Snipes. G.J. Newton, A.F. Eidson, J.L. Mauderly, and B.B. Boecker. 1990. Predicted and observed eraly effects of combined alpha and beta lung irradiation. Health Phys. 59:791-805.

Shubik, P., ed. 1978. Special Problems with Carcinogenicity Protocols. Toxicology Forum.

Sipes, I.G., and A.J. Gandolfi. 1991. Biotransformation of toxicants. Pp. 88-126 in Caserett and Doull's Toxicology, M.O. Amdur, J. Doull, and C.D. Klaassen, eds. 4th ed. New York: Pergamon Press.

Sontag, J.M., N.P. Page, and U. Saffiotti. 1976. Guidelines for Carcinogen Bioassay in Small Rodents. Carcinog. Tech. Rep. Series 1. DHEW Publication (NIH 76-801). Bethesda, Maryland.: National Cancer Institute.

Stephenson, R.P. 1956. A modification of receptor theory. Brit. J. Pharmacol. 11:379–.

Tennant, R.W., M.R. Elwell, J.W. Spalding, and R.A. Griesemer. 1991. Evidence that toxic injury is not always associated with induction of chemical carcinogenesis. Mol. Carcinogen. 4:420-440.

Tischler, A.S., R.A. Delellis, G. Nunnemacher, and H.J. Wolfe. 1988. Acute stimulation of chromaffin cell proliferation in the adult rat adrenal medulla. Lab. Invest. 58:733-735.

Tischler, A.S., L.A. Ruzicka, S.R. Donahue, and R.A. Delellis. 1989. Chromaffin cell proliferation in the adult rat adrenal medulla. Int. J. Devel. Neurosci. 7:439-448.

Tischler, A.S., R.M. McClain, H. Childers, and J. Downing. 1991. Neurogenic signals regulate chromaffin cell proliferation and mediate the mitogenic effects of reserpine in the rat adrenal medulla. Lab. Invest. 65:374-376.

Turnbull, G.J., P.N. Lee, and F.J.C. Roe. 1985. Relationship of body weight to longevity and to risk of development of nephropathy and neoplasia in Sprague-Dawley Rats. Food Chem. Toxicol. 23:355–362.

Umbreit, T.H., and M.A. Gallo. 1988. Physiological implications of estrogen receptor modulation by 2,3,7,8-tetrachlorodibenzo-*p*-dioxin. Toxicol. Lett. 42:5–14.

Wada, S., M. Hirose, S. Takahashi, S. Okazaki, and N. Ito. 1990. *Para*-methoxyphenol strongly stimulates cell proliferation in the rat forestomach but is not a promoter of rat forestomach carcinogenesis. Carcinogenesis 11:1891–1894.

Ward, J.M., N. Konishi, and B.A. Diwan. 1990. Renal tubular cell or hepatocyte hyperplasia is not associated with tumor promotion by di(2-ethylhexyl)phthalate in B6C3F1 mice after transplacental initiation with *N*-nitrosoethylurea. Exp. Pathol. 40:125–138.

Whitlock, J.P. 1986. The regulation of cytochrome P-450 expression. Annu. Rev. Pharmacol. Toxicol. 26:333–369.

Williams, G.M., ed. 1988. Sweeteners: Health Effects. Princeton, N.J.: Princeton Scientific Publishing Co., Inc.

Zeise, L., R. Wilson, and E. Crouch. 1984. Use of acute toxicity to estimate carcinogenic risk. Risk Anal. 4(3):187–199.

Zeise, L., E.A.C. Crouch, and R. Wilson. 1985. Reply to comments: On the relationship of toxicity and carcinogenicity. Risk Anal. 5(4):265–270.

Zeise, L., E.A.C. Crouch, and R. Wilson. 1986. A possible relationship between toxicity and carcinogenicity. J. Amer. Coll. Toxicol. 5(2):137–151.

.

Appendix A

Workshop Summary

MAXIMUM TOLERATED DOSE:
IMPLICATIONS FOR RISK ASSESSMENT

INTRODUCTION

This report summarizes the discussions at a workshop held by the Committee on Risk Assessment Methodology on September 6, 1990, in Washington, DC. An agenda and a list of presenters, discussants, and other participants are appear in Appendixes D and E.

BACKGROUND

Current testing for carcinogenicity in laboratory animals involves testing both sexes of rats and mice for 2 years (nearly a lifetime) at an estimate of the maximum tolerated dose (MTD) and usually at one or more lower doses. The MTD is defined generally as the highest dose of the test agent that is predicted not to alter the animals' longevity or growth because of noncancer effects. The MTD thus varies inversely with the toxicity of a chemical.

A number of researchers have investigated correlations between the MTD and various measures of carcinogenic potency. Some have concluded that the correlations have a biologic basis and might be indicating something general about mechanisms of carcinogenesis. Others have suggested that the correlations result from the selection of data and the methods of analysis themselves and that they do not have a biologic basis. The main intent of this workshop was to investigate the nature of the correlations and what could be concluded from them. The workshop also addressed related correlations between the LD_{50} (the dose estimated to kill 50% of animals) and measures of carcinogenic potency and between measures of carcinogenic potency in different species. To have a firm basis for its deliberations, the committee solicited information on the definitions and methods of determining the various measures of toxicity and carcinogenic potency involved in the correlations.

In the last few years, testing at the MTD has been criticized as providing too sensitive a screen for carcinogenicity. Approximately half the materials tested to date in studies using the MTD as one of the doses tested have shown statistically significant increases in cancer incidence in one or more of the four sex-species groups usually tested. It has been suggested that part of the putative over sensitivity occurs because testing at the MTD induces carcinogenesis by mechanisms that are likely not to be operable at lower doses. One suggested mechanism is systemic toxicity, which leads to excess cellular proliferation and ultimately to the development of cancer.

The workshop addressed those criticisms of testing at the MTD. Evidence of various mechanisms of cancer induction at the MTD and their importance at lower doses was presented. The rationale for testing at the MTD and methods for estimating the MTD were discussed. The proper way to interpret results obtained at the MTD was also discussed, as well as some alternative methods for selecting the highest dose for a cancer bioassay.

DEFINING AND DETERMINING THE MTD

Eugene McConnell, the introductory speaker, described how the MTD as currently used is determined (McConnell, 1989):

The NCI publication by Sontag et al. in 1976, entitled *Guidelines for Carcinogen Bioassay in Small Rodents*, became a standard reference. This publication is known particularly for its definition of an MTD: "The MTD is defined as the highest dose of the test agent during the chronic study that can be predicted not to alter the animals' longevity from effects other than carcinogenicity." The authors further stated that the dose should be one that "causes no more than a 10% weight decrement, as compared to the appropriate control groups, and does not produce mortality, clinical signs of toxicity, or pathologic lesions (other than those that may be related to a neoplastic response) that would be predicted [in the chronic study] to shorten an animal's natural life span." Since that time, Sontag et al.'s definition has been restated and redefined by several groups and authors but remains essentially the same. However, in practice there is a different emphasis. In using these guidelines the primary parameter currently for selecting the MTD is the histopathological appearance, with effects on weight gain being a secondary consideration.

Dr. McConnell emphasized that the estimated MTD (EMTD) is selected on the basis of a 90-day or other prechronic test and involves scientific judgment applied to the information available at the end of the test period. Whether the true MTD was administered can be evaluated only after the bioassay has been conducted.

John Emmerson, speaking from the perspective of pharmaceutical research, stated that the current long-term rodent carcinogenicity studies lack five of the properties that contribute to precision and reliability of the classical new-drug bioassay:

1. A specific biologic end point, whose attributes have been determined experimentally.

• In the carcinogenicity bioassay, one can quantify a response, but the type of tissue affected, the type of tumor seen, the induction time, etc., are not known at the beginning of the study.

• The bioassay is basically a discovery process, which in any other experimental procedure would provide the first data that would permit an investigator to ask good questions and formulate hypotheses for testing in followup studies.

2. A test system in which the investigator can measure a graded response to increasing doses.

• The potential for one to obtain good dose-response data is present; without foreknowledge of the potency of the test substance, the potential is not often realized. Most frequently, a response is observed only at the high dose, the MTD.

3.   A reference standard that permits comparison to the test substance and quantification of potency.

• This is nonexistent for virtually all new substances that are assayed.

• Attempts to define potency are usually feeble and wholly unsatisfactory, because of the tenuous bridges between the effects observed with chemically unrelated substances tested at different times in assays that fail to provide adequate dose-response data.

4.   An experimental design that can be readily tested to confirm the sensitivity of the assay and rule out the presence of unexpected or unknown variables.

• The length of the bioassay precludes ready affirmation of the sensitivity of the animals to the end point.
• Whether there are unexpected or unknown variables that affect the outcome is recognized only in retrospect.

5.   The ability to define potency in units that have reliable application to humans.

• There is no opportunity for planned, experimental confirmation in humans.

Dr. Emmerson proposed that, in addition to the current definition of MTD, there be added the proviso that the dose be selected with a reasonable assurance that the kinetics of systemic exposure and the physiologic response to treatment are quantitatively proportionate and qualitatively similar to those observed in animals to be given lower doses.
Ian Munro suggested that the MTD should be a dose that:

1.   Is adequate to characterize the chronic toxicity of the chemical without inducing overt toxicity that leads to untimely death from effects that would preclude tumor development.

2.   Does not induce gross disturbances in organ function (as determined by clinical and biochemical methods) that would produce a physiologic state incompatible with normal clinical function.

3.   Is chosen with a full understanding of the pharmacokinetics and metabolic profile in relation to dose, so that one knows in advance that rate-limiting mechanisms or qualitative changes in metabolism can be at play in development (this is important in deciphering whether tumorigenic effects are due to secondary mechanisms).

4.   Does not exceed a dose that produces alterations in nutrient intake or use, lest it be of limited relevance to humans.

The first and third points would not lead to estimation of an EMTD different from that obtained with the current approach. The stipulations in the third point advise that the bioassay be initiated with more information gathered in advance.

In reply to a specific question, Dr. Munro said, "I don't think any of us are saying we should not use high dose in testing."

Daniel Krewski presented the major evidence concerning the correlation between EMTD (or, more accurately, highest dose tested, HDT) and an inverse measure of carcinogenic potency, TD_{50} (the 50% excess-tumor-response dose). The more potent the carcinogen, the lower the expected TD_{50}—i.e., more potent carcinogens produce tumors at lower doses than less potent carcinogens. Dr. Krewski's results are presented in Appendix F. Using data derived from the Carcinogenic Potency Data Bank (CPDB) developed by L. Gold and associates, Krewski et al. related the TD_{50}s and the HDT for 191 compounds. The 191 compounds "were selected to satisfy a number of criteria, including the requirements that the experiment have at least two doses in addition to the controls and demonstrate clear evidence of carcinogenicity." Dr. Krewski pointed out that, in some sense, whatever correlations he found would likely be understatements, inasmuch as both the TD_{50} and the HDT (presumed MTD) would be subject to experimental error. He described a "shrinkage" technique to reduce the overdispersion. However, this technique had only a small effect.

Several dose-response models were used to estimate the TD_{50} and the Pearson coefficients of correlation between log TD_{50} and log HDT were computed. The coefficients are given in Table 1 of Appendix F.

Because of the nature of the limitations on the data, if one makes some reasonable assumptions (e.g., the HDT for different carcinogens follows a log-normal distribution, and the TD_{50} is uniformly distributed about the HDT "within the limits calculated by Bernstein"), the correlation between log TD_{50} and log MTD is 0.924. That high (theoretical) correlation suggests that the TD_{50} (hence, the carcinogenic potency for a material established as a carcinogen) could be predicted by the MTD. Krewski et al. cited further investigation on whether the salmonella-microsome assay might be used to predict carcinogenic potency. A statistically significant positive correlation of r = 0.48 was reported and implies that the overall scatter was so great as "to preclude the use of the Ames [salmonella-microsome] assay as a quantitative predictor of carcinogenic potency."

Krewski's conclusion was that preliminary estimates of the low-dose cancer risk can be based on an estimate of the MTD. Citing Gaylor (1989), he reported that dividing the MTD by a factor of 380,000 will approximate the 10^{-6}-risk dose if the linearized multistage model is used—without determining whether the material is a carcinogen. On the general issue of TD_{50}-MTD correlations, he says that "correlations between MTD and the TD_{50} occur as a result of the narrow range of possible potency values within a single experiment in relation to the wide variation observed in the potency of chemical carcinogens." He notes that fact in relation to "suggestions that the observed correlation between the MTD and the TD_{50} may simply be an artifact of the experiment designs currently used in carcinogen bioassay." He further stated that "this does not imply that estimates of carcinogenic potency based on bioassay data are not meaningful, but does demonstrate that both the TD_{50} and q_1^* [the upper 95% confidence limit linear term in the linearized multistage dose-response curve] represent relatively crude indicators of risk."

Krewski noted that measures other than the EMTD could also be used as predictors of carcinogenic potency—for example, acute toxicity—as measured by the LD_{50}. He also reviewed the data on the correlations of carcinogenic potency among different species and remarked that, in view of the high correlations between potency and the HDT and the wide range of carcinogenic potency, one should expect to find high correlations between separate species. The predictability from one species to another is "not within a factor of one and a half or two, but may be tenfold in either direction," which Krewski considers rather good.

Krewski et al. did not reach any firm conclusions. The two ends of the spectrum that they saw were: (1) "Because of all these correlations, some of which may be artifactual, we don't have a good instrument . . . to assess the human cancer risk." (2) "The limits we are currently using are . . . statistically as consistent as you can actually get with the experimental data . . . . From that point of view they are reasonable." (He added that "it seems that additional information beyond that contained in traditional experiments will be required . . . . More sensitive indicators of effects at very low doses . . . may also serve to provide improved estimates of risk in the future.")

The discussants who followed Dr. Krewski were Kenny Crump, Lauren Zeise, Thomas Starr, and Edmund Crouch.

Dr. Crump reported on the correlations between the laboratory animal data and the potency (as measured by a TD_{25} dose in humans for 23 confirmed carcinogens) (Allen et al., 1988). He and his colleagues found a high correlation of about 0.8. (Different methods of analysis involving different assumptions led to slightly different correlations.) His data showed, qualitatively, "that chemicals that are more highly carcinogenic in animals tend to be also more highly carcinogenic in humans," despite the fact that humans are rarely exposed to a human equivalent of a laboratory MTD, or for a lifetime of exposure. He added that "we did not find . . . that the estimates currently being made tended to grossly overestimate or underestimate the risks actually estimated directly from the . . . data."

Dr. Zeise reported on her work with Crouch and Wilson related to the correlations, asking whether the correlations were "real" or artifactual and whether they could be used to predict carcinogenic potency (Zeise et al., 1984; 1985; 1986). She reported that materials tested at doses that caused weight depression early were more likely to be reported as carcinogens than materials that did not. In that regard, she referred to the work of Haseman (1985), who found that about 85% of the carcinogens examined yielded no evidence of cell proliferation in at least one of the tissues where cancers were found.

With respect to the departures from the relationship between HDT and potency, she found few materials that clearly resulted in low toxicity and high cancer potency. (Such materials would be the so-called super carcinogens.) Dr. Zeise did note that several materials had low toxicity and produced cancers after a very short exposure; they might therefore be considered supercarcinogens of another kind and had been excluded from her study. Among materials of that type were three benzidine dyes that produced cancers in 13 weeks—after which the experiments were terminated.

Dr. Zeise concluded that the correlations were not completely artifactual and that some analyses on individual animals, considering time to tumor, might improve the estimates. She noted, however, that the data did not answer the question of whether "toxicity is in fact causing cancer."

Dr. Starr recalled an earlier paper of his with Rieth (Rieth and Starr, 1989a) and, by way of summary, reiterated the conclusion of that paper: "We hold the opinion that the chronic rodent bioassay in and of itself is altogether inadequate as the data source for estimating the risk to humans from exposure to carcinogenic chemicals." He also quoted Ames and Gold (1990): "Thus, without studies of the mechanism of carcinogenesis, the fact that a chemical is a carcinogen at the MTD provides no information about low dose risk to humans." He also reported on his more recent work (Rieth and Starr, 1989b) looking at the upper bounds of estimates of carcinogenic potency (by looking at lower bounds of $TD_{50}s$) and comparing materials that were carcinogenic in both rats and mice, in one species only, or in neither. Dr. Starr found that studies that yielded no evidence of carcinogenicity still yielded evidence of correlation ("nearly as good") between upper-bound estimates of potency and the HDT. The correlation he reported for the so-called negative-negative materials was 0.88, the highest of the four correlations he computed. That and related computations led him to conclude that "it's my opinion, at least, that this business is not giving us much information that is useful at all in quantifying low-dose risk." Finally, he objected to Krewski's word "measure" and stated that he preferred to use the phrase "crude estimate," noting, among other things, that all estimates are model-dependent.

Dr. Starr's proposals to escape the problems he discussed involve going from an administered dose to a target-tissue dose and moving to a model more like the "two-stage growth-death model of Moolgavkar"

(Moolgavkar and Venzon, 1979; Moolgavkar and Knudson, 1981). They should also involve looking at cell turnover rates in the normal cell compartment and in the initiated-cell compartment.

Dr. Crouch expressed concern about imposing constraints on the data because of mathematical needs or imposing limits in estimation that are not imposed by nature ("nobody imposed that constraint on the animals"). He expressed his belief that the correlations were real and important: "The reason that you're getting the correlations is that nature is telling you something." He also proposed several alternatives, including that the test material (at the doses tested) directly caused DNA damage in cells and that such damage might lead to both acute toxic and carcinogenic effects. However, risk assessment does not require knowledge of causality, but only knowledge of correlations.

The afternoon session was entitled "What are Bioassays Conducted at the MTD Telling Us?" The presenter was Bruce Ames, substituting for his colleague Lois Gold. Dr. Ames's discussion touched on evolutionary issues, cancer as a disease of old age, the mutagenic activity of oxygen radicals, and responses to infection. That led into a discussion of DNA damage and (somatic) mutation, and he noted that "it's hard to mutate a cell unless its dividing." From that and related arguments, Dr. Ames developed the idea that promotion is essentially related to cell division and that what stimulates cell division increases the likelihood of cancer development. He cited the work by Henderson and Preston-Martin (1990), who associated several human cancers with agents "causing a lot of cell proliferation."

Dr. Ames described how the CPDB was developed, and from there he moved into an argument about the importance of so-called natural carcinogens, of which slightly fewer than half are positive (i.e., are carcinogenic) in one or more species-sex groups. He argued that that is an extremely important finding, because "almost all the chemicals in the world are natural." He pointed out that plant breeders are breeding plants to be insect-resistant, and some of the natural insecticides developed (or increased) in breeding programs might act as carcinogens for humans, although the toxic chemicals in plants tend to be species-specific. He remarked that "we are living in a world of toxic chemicals which come from these plants." In addition, he noted the likely presence of many plant anticarcinogens.

Eugene McConnell, commenting on Dr. Ames's discussion of the place of cell proliferation in carcinogenesis said that "having looked at between 100 and 200 chemicals in animal bioassays [I find that] . . . many of these [carcinogenic] chemicals show cell proliferation in organs where the tumors are seen. . . . I also note that . . . many chemicals that also cause cell proliferation . . . are not carcinogens." In reply to a comment by Richard Reitz on the effects of applying risk assessment methods to materials in a common diet, Dr. Ames suggested testing a "random group of nature's pesticides." Dr. Ames (replying to a question by Jill Snowden) again raised the issue of the presence of anticarcinogens in fruits and vegetables.

Returning to the MTD issue, Miriam Davis noted that "90% of chemicals that were carcinogenic at high doses were also producing . . . tumors at lower doses." That was followed by a discussion of the results of testing at doses lower than the MTD—with some suggestions for testing more food chemicals, but with no recommendations about the number of animals needed to retain tests of satisfactory power. Dr. Ames's major comment was that "when you have a high dose, it's hard to go to a low dose . . . We have to learn more about mechanisms to predict . . . [whether something is] a carcinogen."

Michael Gallo gave a history of carcinogenicity testing, including the considerations that enter into the selection by the National Toxicology Program of materials to test. He characterized the current studies as "excellent toxicology" but noted that they were not designed for risk analysis. Risk analysis requires more information than the current bioassays present. Dr. Gallo recommended short-term testing at toxic doses and "then . . . back[ing] down the dose-response curve and defin[ing] the shape of that . . . response curve."

David Gaylor discussed the statistical issues and the data that could and could not be added by modifications in the design of the current bioassays (e.g., by adding more low doses). He noted that "persons who do risk estimation [do not] believe . . . that the number we come up with is in any sense a precise number. . . . But, apart from the low-dose extrapolation, the uncertainty in the data seems to be on the order of a maximum of about 100." That being the case, "bioassays are perhaps getting us in the right ballpark and certainly can be used to rank carcinogens." He noted that low-dose extrapolation is more affected by the results at the low doses than at the high dose. Furthermore, the higher the background rate of cancer, the more likely that there will be a linear term in the fitted dose-response model. Dr. Gaylor invoked the idea of endogenous factors that produce tumors without the addition of chemicals as an argument to demonstrate that low-dose effects added to background should increase tumor rates.

In the ensuing discussion stimulated by prepared remarks of Drs. Munro, Wilson, and Engler, issues were raised about the place (and development) of more appropriate biologic models, the relative potency of mutagenic versus nonmutagenic carcinogens, the classification and ranking of carcinogens (based on weight of evidence, rather than potency), and the application of bioassay results to prevent (putative) risks lower than can be measured epidemiologically or with bioassays. The point was made that regulators need be concerned about the expected human exposure (dose) as much as or more than about an absolute measure of potency. A highly potent material to which no one is, or can be, exposed poses no risk. Finally, data were introduced to show the potential for predicting carcinogenicity (for chemicals in well-defined specific classes) by using data from, for example, the salmonella short-term assays. It was noted that, operationally, "60% of the substances that have come to NTP [for testing have come] because of the suspicion of carcinogenicity."

Options for performing bioassays and for using bioassay data in the identification of human carcinogens were extensively discussed. The options are incorporated in a modified form in the main body of this report.

Appendix B

Organizing Subcommittee

Kenny S. Crump (Chairman)
Clement Associates, Inc.
Ruston, LA

Paul T. Bailey
Mobil Oil Corporation
Princeton, NJ

Michael A. Gallo
Robert Wood Johnson Medical
 School
University of Medicine and
 Dentistry of New Jersey
Piscataway, NJ

Richard A. Griesemer
National Institute of Environ-
 mental Health Sciences
National Toxicology Program
Research Triangle Park, NC

Dale Hattis
Clark University
Worcester, MA

Rogene Henderson
Lovelace Biomedical and Environmental Research Institute
Albuquerque, NM

Daniel Krewski
Health and Welfare Canada
Ottawa, Ontario
Canada

Donald R. Mattison
University of Pittsburgh
Graduate School of Public
 Health
Pittsburgh, PA

Ian Nisbet
I.C.T. Nisbet & Company,
 Inc.
Lincoln, MA

Richard H. Reitz
The Dow Chemical Company
Midland, MI

Appendix C

Federal Liaison Group

Dr. Deborah Barsotti
Agency for Toxic Substances
   and Disease Registry
Atlanta, GA

Dr. James Beall
U.S. Department of Energy
Washington, DC

Dr. William Cibulas
Agency for Toxic Substances
   and Disease Registry
Atlanta, GA

Dr. Murray S. Cohn
Consumer Product Safety
   Commission
Bethesda, MD

Dr. Joseph Cotruvo
U.S. Environmental Protection
   Agency
Washington, DC

Dr. William H. Farland
U.S. Environmental Protection
   Agency
Washington, DC

Dr. Herman Gibb
U.S. Environmental Protection
   Agency
Washington, DC

Dr. Walter H. Glinsmann
U.S. Food and Drug Adminis-
   tration
Washington, DC

Dr. Bryan D. Hardin
National Institute for Occupa-
   tional Safety and Health
Cincinnati, OH

Dr. Peter Infante
U.S. Department of Labor/
   OSHA
Washington, DC

Dr. Ronald J. Lorentzen
U.S. Food and Drug Adminis-
    tration
Washington, DC

Dr. Edward Ohanian
U.S. Environmental Protection
    Agency
Washington, DC

Mr. Richard Orr
U.S. Department of Agricul-
    ture
Hyattville, MD

Dr. Richard Parry, Jr.
U.S. Department of Agricul-
    ture
Beltsville, MD

Dr. Dorothy Patton
U.S. Environmental Protection
    Agency
Washington, DC

Dr. Peter Preuss
U.S. Environmental Protection
    Agency
Washington, DC

Dr. Lorenz R. Rhomberg
U.S. Environmental Protection
    Agency
Washington, DC

Dr. Matthew H. Royer
U.S. Department of Agricul-
    ture
Hyattsville, MD

Dr. Lilly Sanathanan
U.S. Food and Drug Adminis-
    tration
Rockville, MD

Dr. Robert Scheuplein
U.S. Food and Drug Adminis-
    tration
Washington, DC

Ms. Janet A. Springer
U.S. Food and Drug Adminis-
    tration
Washington, DC

Dr. Bruce V. Stadel
U.S. Food and Drug Adminis-
    tration
Rockville, MD

Dr. Leslie T. Stayner
National Institute for Occupa-
    tional Safety and Health
Cincinnati, OH

Mr. Michael T. Werner, JD
U.S. Department of Agricul-
    ture
Hyattsville, MD

Appendix D

Workshop Program

Thursday, September 6, 1990

Lecture Room, National Academy of Sciences
2101 Constitution Ave., NW
Washington, DC

9:00   **Introduction and Objectives:** Kenny Crump, Clement Associates; Chairman

9:15   **Definition and Application of MTD**
      Presenter: Eugene McConnell, Raleigh, NC
9:45   Discussants: John Emmerson, Eli Lilly and Co.
                Ian Munro, University of Guelph
10:05 Questions and Comments

10:15 Break

10:30 **Correlations between the MTD and Measures of Carcinogenic Potency: Implications for Risk Assessment**
      Presenter: Daniel Krewski, Health and Welfare Canada
11:15 Discussants: Kenny Crump, Clement Associates
      Lauren Zeise, California Department of Health Services
      Thomas Starr, Environ Corp.

Edmund Crouch, Cambridge Environmental, Inc.

12:00   Questions and Comments

12:30 Lunch

2:00   **What are Bioassays Conducted at the MTD Telling Us?**
        Presenter: Bruce Ames, University of California, Berkeley
2:45   Discussants: Michael Gallo, University of Medicine and
                      Dentistry of New Jersey
                      David Gaylor, National Center for Toxicologi-
                      cal Research
3:15   Questions and Comments

3:30   Break

3:45   **General Discussion**
        Leader: Kenny Crump, Clement Associates
        Issues:
          • The basis of the observed correlation between the
            MTD and measures of carcinogenic potency
          • The possible influence on biological processes of
            dosing at this level
          • The implications for the design of rodent bioassays
          • The implications for carcinogenic risk assessment

5:00   Adjourn

Appendix E

Workshop Attendees

Dr. Philip H. Abelson
AAAS
Washington, DC

Dr. Bruce Ames
University of California, Berkeley
Berkeley, CA

Dr. Alexander Apostolou
U.S. Food and Drug Administration
Rockville, MD

Dr. Gail T. Arce
Alochem North America
Agrochemicals Division
Philadelphia, PA

Dr. Debra Aub
U.S. Food and Drug Administration
Rockville, MD

Dr. Karl Baetcke
U.S. Environmental Protection
Agency
Washington, DC

Dr. Paul T. Bailey
Mobil Oil Corporation
Princeton, NJ

Mr. Robert C. Barnard
Cleary, Gottlieb, Steen and
Hamilton
Washington, DC

Dr. Karam Batra
U.S. Food and Drug Administration
Rockville, MD

Mr. Gregory Bendlin
Cameron & Hornbostel
Washington, DC

Dr. Jeff Beaubier
U.S. Environmental Protection
   Agency
Washington, DC

Ms. Michele Beguhn
ICI Americas
Wilmington, DE

Dr. Judith Bellin
Washington, DC

Mr. Keith Belton
American Chemical Society
Washington, DC

Dr. Richard Belzer
Office of Management and
   Budget
Washington, DC

David Bergsten
U.S. Department of Agricul-
   ture
Hyattsville, MD

Dr. Sanford W. Bigelow
National Research Council
Washington, DC

Ms. Nathalie Blackwell
Ketchum Communications
Washington, DC

Dr. Charles Blank
American Petroleum Institute
Washington, DC

Dr. Charles Breckenridge
CIBA GEIGY Corporation
Greensboro, NC

Dr. E. Anne Brown
U.S. Department of Agricul-
   ture
Washington, DC

Dr. Robert Brown
U.S. Food and Drug Adminis-
   tration
Washington, DC

Dr. Gary Burin
Falls Church, VA

Dr. William Burman
U.S. Environmental Protection
   Agency
Washington, DC

Dr. Dorothy Canter
National Institute of Environ-
   mental Health Sciences
Bethesda, MD

Dr. Raymond Cardona
Uniroyal Chemical Company
Bethany, CT

Dr. Charles J. Carr
Scientific Information Associ-
   ates
Columbia, MD

Dr. Chia Chen
U.S. Department of Labor/
OSHA
Washington, DC

Dr. Paul Chin
U.S. Environmental Protection
Agency
Washington, DC

Dr. Arthur Chiu
U.S. Environmental Protection
Agency
Washington, DC

Dr. Young S. Choi
U.S. Food and Drug Administration
Rockville, MD

Dr. Misoon Chun
Cabin John, MD

Mr. Dan Charles
New Scientist
Washington, DC

Dr. William Cibulas
Agency for Toxic Substances
and Disease Registry
Atlanta, GA

Mr. Steve Clapp
Food Chemical News
Washington, DC

Mr. David Clark
Inside EPA Weekly Report
Arlington, VA

Dr. Ann Clevenger
U.S. Environmental Protection
Agency
Washington, DC

Dr. Rory Conolly
Chemical Industrial Institute of
Toxicology
Research Triangle Park, NC

Dr. Joseph Contrera
U.S. Food and Drug Administration
Rockville, MD

Mr. Robert A. Coppock
National Research Council
Washington, DC

Dr. Edmund Crouch
Cambridge Environmental
Cambridge, MA

Dr. Kenny S. Crump
Clement Associates, Inc.
Ruston, LA

Dr. Cynthia Cunard
U.S. Food and Drug Administration
Rockville, MD

Dr. Miriam Davis
National Institute of Environ-
    mental Health Sciences/NTP
National Institutes of Health
Bethesda, MD

Dr. Richard A. Davis
American Cyanamid Company
Wayne, NJ

Ms. Kelly Day
U.S. Department of Agricul-
    ture
Washington, DC

Dr. Kerry Dearfield
U.S. Environmental Protection
    Agency
Washington, DC

Mr. Charles L. Divan
Environmental Documentation
Hyattsville, MD

Dr. John Doherty
U.S. Environmental Protection
    Agency
Washington, DC

Dr. John Doull
University of Kansas Medical
    Center
Kansas City, KS

Mr. Robert Drew
American Petroleum Institute
Washington, DC

Dr. Julie Du
U.S. Environmental Protection
    Agency
Washington, DC

Dr. Ronald J. Dutton
International Life Sciences
    Institute
Washington, DC

Dr. Gerard Egan
Exxon Biomedical Sciences,
    Inc.
East Millstone, NJ

Dr. Reto Engler
U.S. Environmental Protection
    Agency
Washington, DC

Dr. John L. Emmerson
Lilly Research Laboratories
Eli Lilly and Company
Greenfield, IN

Dr. William H. Farland
U.S. Environmental Protection
    Agency
Washington, DC

Mr. Conner M. Fay
Clairol, Inc.
New York, NY

Dr. Penelope Fenner-Crisp
U.S. Environmental Protection
    Agency
Washington, DC

Ms. Bernice Fisher
U.S. Environmental Protection
   Agency
Washington, DC

Dr. Glenna Fitzgerald
Silver Spring, MD

Ms. Caroline Freeman
U.S. Department of Labor/
   OSHA
Washington, DC

Dr. Tom Fuhremann
Monsanto Agricultural Compa-
   ny
St. Louis, MO

Ms. Carolyn Fulco
Department of Human and
   Health Services
Washington, DC

Dr. Victor Fung
National Institute of Environ-
   mental Health Sciences
Bethesda, MD

Dr. Wing Fung
Unilever Research
Edgewater, NJ

Dr. Michael A. Gallo
Robert Wood Johnson Medical
   School
University of Medicine and
   Dentistry of New Jersey
Piscataway, NJ

Mr. Hank Gardner
U.S. Army
Frederick, MD

Dr. David Gaylor
National Center for Toxicolog-
   ical Research
Jefferson, AR

Dr. Herman Gibb
U.S. Environmental Protection
   Agency
Washington, DC

Dr. Walter H. Glinsmann
U.S. Food and Drug Adminis-
   tration
Washington, DC

Dr. Anwar Goheer
U.S. Food and Drug Adminis-
   tration
Rockville, MD

Dr. Bernard D. Goldstein
Robert Wood Johnson Medical
   School
University of Medicine and
   Dentistry of New Jersey
Piscataway, NJ

Dr. Thomas Goldsworthy
Chemical Industrial Institute of
   Toxicology
Research Triangle Park, NC

Dr. Michael Gough
Center for Risk Management
Resources for the Future
Washington, DC

Dr. Martin Greene
U.S. Food and Drug Adminis-
    tration
Rockville, MD

Dr. Richard A. Griesemer
National Institute of Environ-
    mental Health Sciences
Research Triangle Park, NC

Dr. Stanley B. Gross
U.S. Environmental Protection
    Agency
Washington, DC

Ms. Elizabeth Grossman
U.S. Department of Labor/
    OSHA
Washington, DC

Dr. William E. Halperin
National Institute for Occupa-
    tional Safety and Health
Cincinnati, OH

Dr. Bryan D. Hardin
National Institute for Occupa-
    tional Safety and Health
Cincinnati, OH

Dr. Jane Harris
American Cyanamid Company
Princeton, NJ

Dr. Dale Hattis
Center for Technology, Policy,
    and Industrial Development
Massachusetts Institute of
    Technology
Cambridge, MA

Mr. Fred Hauchman
U.S. Environmental Protection
    Agency
Research Triangle Park, NC

Dr. Rogene Henderson
Lovelace Biomedical and Envi-
    ronmental Research Institute
Albuquerque, NM

Dr. Bernhard Hildebrand
BASF AG
Ludwigshafen
West Germany

Dr. Fred D. Hoerger
The Dow Chemical Company
Midland, MI

Dr. Karen Hogan
U.S. Environmental Protection
    Agency
Washington, DC

Mr. Stewart Holm
Halogenated Solvents Industry
    Alliance
Washington, DC

Dr. Charlie Hremat
U.S. Environmental Protection
Agency
Washington, DC

Dr. James Huff
National Institute of Environ-
mental Health Sciences
Research Triangle Park, NC

Dr. Donald Hughes
Procter & Gamble Company
Cincinnati, OH

Dr. Peter Infante
U.S. Department of Labor/
OSHA
Washington, DC

Dr. Carol Jones
U.S. Department of Labor/
OSHA
Washington, DC

Dr. Dennis E. Jones
Agency for Toxic Substances
and Disease Registry
Atlanta, GA

Dr. Alexander Jordan
U.S. Food and Drug Adminis-
tration
Rockville, MD

Dr. A. Michael Kaplan
Haskell Labortory
E.I. duPont de Nemours &
Co.
Newark, DE

Dr. Myra Karstadt
National Institue of Environ-
mental Health Sciences
Bethesda, MD

Mr. William Kelly
Federal Focus, Inc.
Washington, DC

Dr. Krishan Khanna
U.S. Environmental Protection
Agency
Washington, DC

Ms. Leslie J. King
Chemical Manufacturers Asso-
ciation
Washington, DC

Mr. William A. Kittrell
Karch & Associates, Inc.
Washington, DC

Ms. Gina Kolata
New York Times
New York, NY

Dr. Albert C. Kolbye
Kolbye Associates
Bethesda, MD

Ms. Christine Kopral
U.S. Food and Drug Adminis-
  tration
Washington, DC

Dr. Richard Krasula
Abbott Laboratories
Abbott Park, IL

Dr. Daniel Krewski
Health and Welfare Canada
Ottawa, Ontario
Canada

Dr. Betsy Kuhn
U.S. Department of Agricul-
  ture
Washington, DC

Dr. David Lai
U.S. Environmental Protection
  Agency
Washington, DC

Dr. Brian P. Leaderer
John B. Pierce Foundation
Yale University School of
  Medicine
New Haven, CT

Mr. Arnold H. Leibowitz
Cameron & Hornbostel
Washington, DC

Ms. Linda V. Leonard
National Research Council
Washington, DC

Dr. David Lien
U.S. Environmental Protection
  Agency
Washington, DC

Dr. Bertam Litt
Litt Associates
Washington, DC

Dr. Janice Longstreth
Battelle Pacific NW
Washington, DC

Dr. Ronald J. Lorentzen
U.S. Food and Drug Adminis-
  tration
Washington, DC

Dr. Carol Maczka
Clement Associates
Fairfax, VA

Dr. Amal Mahfouz
U.S. Environmental Protection
  Agency
Washington, DC

Dr. Elizabeth Margosches
U.S. Environmental Protection
  Agency
Washington, DC

Dr. Donald Mattison
Graduate School of Public
  Health
University of Pittsburgh
Pittsburgh, PA

Dr. John F. McCarthy
National Agricultural Chemicals Association
Washington, DC

Dr. R. Michael McClain
Hoffmann LaRoche, Inc.
Nutley, NJ

Ernest E. McConnell, DVM
Raleigh, NC

Dr. Lawrence McCray
National Research Council
Washington, DC

Dr. James McDermott
Procter and Gamble Company
Cincinnati, OH

Dr. John Melagrana
U.S. Food and Drug Administration
Rockville, MD

Dr. Frances Mielach
U.S. Food and Drug Administration
Rockville, MD

Dr. Franklin E. Mirer
United Auto Workers
Detroit, MI

Dr. Lakshmi Mishra
Consumer Product Safety Commission
Bethesda, MD

Dr. Alastair Monro
Phizer Central Research
Groton, CT

Dr. Ian C. Munro
Canadian Centre for Toxicology
University of Guelph
Guelph, Ontario
CANADA

Mr. Leonard Nessen
U.S. Food and Drug Administration
Rockville, MD

Dr. Daniel W. Nebert
Institute of Environmental Health
University of Cincinnati Medical Center
The Kettering Laboratory
Cincinnati, OH

Dr. Thoa Nguyen
U.S. Environmental Protection Agency
Washington, DC

Dr. Ian Nisbet
I.C.T. Nisbet & Company, Inc.
Lincoln, MA

Dr. D. Warner North
Decision Focus, Inc.
Mountain View, CA

Ms. Carolyn Nunley
ERM
Exton, PA

Dr. Clyde Oberlander
U.S. Food and Drug Administration
Rockville, MD

Dr. Edward Ohanian
U.S. Environmental Protection Agency
Washington, DC

Dr. Gilbert S. Omenn
School of Public Medicine and Community Medicine
University of Washington
Seattle, WA

Mr. Richard Orr
U.S. Department of Agriculture
Hyattville, MD

Dr. Robert Osterberg
U.S. Food and Drug Administration
Rockville, MD

Ms. Carol O'Toole
American Industrial Health Council
Washington, DC

Dr. Richard Parry, Jr.
U.S. Department of Agriculture
Beltsville, MD

Dr. Yogen Patel
U.S. Environmental Protection Agency
Washington, DC

Dr. Dorothy Patton
U.S. Environmental Protection Agency
Washington, DC

Dr. Mary B. Paxton
National Research Council
Washington, DC

Dr. William Pepelko
U.S. Environmental Protection Agency
Washington, DC

Mr. Hugh Pettigrew
U.S. Environmental Protection Agency
Washington, DC

Dr. Peter Preuss
U.S. Environmental Protection Agency
Washington, DC

Dr. Richard H. Reitz
The Dow Chemical Company
Midland, MI

Dr. Esther Rinde
U.S. Environmental Protection Agency
Washington, DC

Major Welford Roberts
U.S. Environmental Protection Agency
Washington, DC

Dr. Barry Rosloff
U.S. Food and Drug Administration
Rockville, MD

Dr. James Rowe
U.S. Environmental Protection Agency
Washington, DC

Dr. Matthew H. Royer
U.S. Department of Agriculture
Hyattsville, MD

Dr. Manfred Ruthsatz
U.S. Food and Drug Administration
Rockville, MD

Dr. Lilly Sanathanan
U.S. Food and Drug Administration
Rockville, MD

Dr. Robert Scheuplein
U.S. Food and Drug Administration
Washington, DC

Dr. Rita Schoeny
U.S. Environmental Protection Agency
Cincinnati, OH

Dr. Loretta Schuman
U.S. Department of Labor/ OSHA
Washington, DC

Dr. Robert L. Sielken, Jr.
Sielken Incorporated
Bryan, TX

Mr. Fred Siskind
U.S. Department of Labor
Washington, DC

Dr. Jerry M. Smith
Rohm & Haas Company
Philadelphia, PA

Dr. Jill Snowdon
United Fresh Fruit and Vegetable Association
Alexandria, VA

Ms. Susan Snyder
American Paper Institute
Washington, DC

Mr. Hugh Spitzer
American Petroleum Institute
Washington, DC

Ms. Anne Sprague
National Research Council
Washington, DC

Ms. Janet A. Springer
U.S. Food and Drug Adminis-
    tration
Washington, DC

Dr. Thomas B. Starr
ENVIRON Corporation
Arlington, VA9

Dr. Leslie T. Stayner
National Institute for Occupa-
    tional Safety and Health
Cincinnati, OH

Dr. Bonnie R. Stern
U.S. Environmental Protection
    Agency
Washington, DC

Ms. Jennifer Sutton
Association of American Medi-
    cal Colleges
Washington, DC

Dr. Alan Taylor
U.S. Food and Drug Adminis-
    tration
Rockville, MD

Ms. Alison Taylor
Harvard School of Public
    Health
Boston, MA

Dr. Richard D. Thomas
National Research Council
Washington, DC

Ms. Sarah Thurin
BNA
Washington, DC

Dr. John J. Tice
Georgia Pacific
Washington, DC

Dr. Lorna C. Totman
Nonprescription Drug Manu-
    facturers Association
Washington, DC

Dr. Curtis C. Travis
Oak Ridge National Laborato-
    ries
Oak Ridge, TN

Dr. Marsha Van Gemert
U.S. Environmental Protection
    Agency
Washington, DC

Dr. Ron Van Mynen
Chemical and Plastics Group
Union Carbide Corporation
Danbury, CT

Dr. Andrea A. Wargo
Agency for Toxic Substances
and Disease Registry
Washington, DC

Dr. Judy Weissinger
U.S. Food and Drug Administration
Rockville, MD

Dr. Bob West
Food, Drug, Chemical
Services
Fairfax, VA

Dr. John Whysner
American Health Foundation
Valhalla, NY

Dr. James Wilson
U.S. Food and Drug Administration
Rockville, MD

Dr. James D. Wilson
Monsanto
St. Louis, MO

Dr. Lucy Yu
U.S. Food and Drug Administration
Rockville, MD

Dr. Lauren Zeise
California Department of
Health Services
Berkeley, CA

Appendix F

Correlation Between Carcinogenic Potency and the Maximum Tolerated Dose: Implications for Risk Assessment

D. Krewski,[1,2] D.W. Gaylor[3], A.P. Soms[4,5] & M. Szyszkowicz[1]

Current practice in carcinogen bioassay calls for exposure of experimental animals at doses up to the maximum tolerated dose (MTD). Such studies have been used to compute measures of carcinogenic potency such as the TD_{50} as well as unit risk factors such as q_1^* for predicting low dose risks. Recent studies have indicated that these measures of carcinogenic potency are highly correlated with the MTD. Carcinogenic potency has also been shown to be correlated with indicators of mutagenicity and toxicity. Correlation of the MTDs for rats and mice implies a corresponding correlation in TD_{50} values for these two species. The implications of these results for cancer risk assessment are examined in light of the large variation in potency among chemicals known to induce tumors in rodents.

1. Introduction

Carcinogen bioassay is an important source of information on the potential carcinogenic effects of chemicals. Current practice involves the exposure of animals at doses up to the maximum tolerated dose

(MTD), defined as that dose which can be administered to rodents over the course of a lifetime without appreciably altering body weight or survival other than as a result of tumor occurrence (Munro, 1977). High doses such as the MTD are used to enhance tumor response rates, thereby increasing the likelihood of observing elevated tumor occurrence rates in a small sample of experimental animals. In this regard, Haseman (1985) has shown that more than two-thirds of the carcinogenic effects detected in feeding studies conducted under the U.S. National Toxicology Program (NTP) would have been missed if the highest dose had been restricted to one-half of the MTD.

The use of such high doses in animal cancer tests has been the subject of considerable debate (cf. McConnell, 1989). In particular, it has been argued that biochemical and physiological distortions occurring at high doses may lead to toxicity-induced carcinogenic effects that might not be expected to occur at lower doses (Carr & Kolbye, 1991; Clayson et al., 1992). Ames & Gold (1990) have suggested that high dose stimulation of mitogenesis will enhance mutagenesis, leading to the identification of rodent carcinogens that may not present a human health risk. Apostolou (1990) questioned the necessity of using the MTD in animal cancer tests on the grounds that many human carcinogens can be identified in animal tests at doses of one-half of the MTD or less.

Suggestions for redefining the high dose to be used in animal cancer tests to circumvent these issues have been made (Apostolou, 1990; Carr & Kolbye, 1991). Clayson et al. (1992) considered such proposals, but recommended retaining the MTD, while recognizing that nongenotoxic carcinogens that appear to be effective in animals only at high doses may not present a risk to humans exposed to much lower doses (cf. Butterworth, 1990). Since the definition of the maximum dose to be used in animal cancer tests is of secondary importance for our present purposes, we make no attempt to resolve this issue here. Instead, the reader is referred to the recent report by the National Research Council (1992), which considers the definition of the maximum dose to be used in detail.

The completion of several hundred bioassays over the past two decades has resulted in the availability of a large data base that may be used in global analyses of bioassay data. Recent analyses have revealed that the MTD is highly correlated with quantitative measures of carcinogenic potency such as the TD_{50} (Bernstein et al., 1985; Reith and Starr, 1989a), defined as the dose that reduces the proportion of tumor free animals by 50% (Peto et al.,1984).

Since the maximum dose tested (MDT) in carcinogen bioassay may not always correspond to the maximum tolerated dose (MTD), we note that it may be more appropriate to claim that carcinogenic potency is correlated with the MDT rather than the MTD. However, since highest dose tested in most studies approximates the MTD, we will not always distinguish between the MDT and the MTD in what follows.

Carcinogenic potency has also been shown to be correlated with various measures of toxicity and mutagenic potential (Travis et al., 1990a). The MTD for rats has also been shown to be correlated with the MTD for mice, for carcinogens that are effective in both species, thereby implying a correlation between the TD_{50} values for these two species (Crouch and Wilson, 1979; Reith and Starr, 1989b).

These meta-analytic results have important implications for carcinogenic risk assessment. The correlation between the MTD and TD_{50} has led to suggestions that the latter measure of carcinogenic potency is simply an artifact of the experimental design specifying the highest dose to be used in the bioassay (Bernstein et al., 1985) and of the use of an essentially linear dose-response model to estimate the TD_{50} (Kodell et al., 1990). The existence of such a correlation has also led to suggestions that preliminary estimates of cancer risk may be derived from the MTD in the absence of carcinogen bioassay data (Gaylor, 1989).

In this paper, we examine these and other issues involved in the use of carcinogen bioassay data for risk assessment purposes. In section 2, we discuss measures of carcinogenic potency proposed in the literature. The reasons for the apparent correlation between the MDT and carcinogenic potency are explored in section 3. The prediction of the TD_{50} on the basis of indicators of subchronic toxicity and genotoxicity is discussed in section 4, along with the calculation of preliminary estimates of cancer risk based on the MTD. Evidence for interspecies correlation in carcinogenic potency is reviewed in section 5. Our conclusions regarding the implications of these results for carcinogenic risk assessment are presented in section 6.

2. Carcinogenic Potency

2.1 Measures of Carcinogenic Potency

Barr (1985) has reviewed a number of proposed measures of carcinogenic potency. Such indices provide a quantitative measure of carcinogenic potential, which may be used to rank the relative potency of different carcinogens. A widely used measure of potency is the TD_{50} proposed by Peto et al. (1984). Application of the TD_{50} in ranking chemical carcinogens has recently been discussed by Woodward et al. (1991); the TD_{50} also represents a primary component of the multifactor ranking scheme proposed by Nesnow (1990). Letting P(d) denote the probability of a tumor occurring in an individual exposed to dose d, the TD_{50} is defined as the dose d that satisfies the equation

$$R(d) \ = \ [P(d)-P(0)]/[1-P(0)] \ = \ 0.5, \qquad (2.1)$$

where R(d) is the extra risk over background at dose d. Thus, the TD_{50} is the dose for which the extra risk is equal to 50% or, equivalently, the dose at which the proportion of tumor-free animals is reduced by one-half.

The TD_{50} may be estimated on the basis of tumor response rates observed in laboratory studies involving a series of increasing dose levels. Sawyer et al. (1984) employ an essentially linear one-stage dose-response model for this purpose, with

$$P(d) \ = \ 1 - \exp\{-(\alpha + \beta d)\} \ . \qquad (2.2)$$

The slope parameter β in this one-hit model is related to the TD_{50} by

$$\beta \ = \ \log_e(2)/TD_{50}, \qquad (2.3)$$

and has been used as a measure of potency by Crouch and Wilson (1981). To accommodate curvature, however, a nonlinear model such as the multi-stage (Armitage, 1985)

$$P(d) \ = \ 1 - \exp\{-(q_0 + q_1 d + \ldots + q_k d^k)\} \qquad (2.4)$$

($q_i \geq 0$) or Weibull (Kodell et al., 1991)

$$P(d) \ = \ 1 - \exp\{-(\alpha + \beta d^k)\} \qquad (2.5)$$

($\alpha, \beta, k > 0$) may be more appropriate. We note that the Weibull model is not being proposed for purposes of low dose risk estimation; rather, it is a relatively simple yet flexible model that allows for curvature in the observable response range.

Another measure of potency, which has been used by the U.S. Environmental Protection Agency (1986), is the estimate of the linear term q_1 in the multi-stage model. Since the extra risk is approximated by

$$R(d) \doteq q_1 d \qquad (2.6)$$

at low doses, the value of q_1 may be used to estimate the risk associated with environmental exposures to a dose d of a carcinogen. In practice, an upper confidence limit q_1^* on the value of q_1 (Crump, 1984a) is used due to the instability of the maximum likelihood estimate of the linear term in the multi-stage model. This application is commonly referred to as the linearized multistage (LMS) model.

Estimates of q_1^* have been criticized on the grounds that they require extrapolation of data well below the experimentally observable tumor response range. The TD_{50}, on the other hand, does not require low dose extrapolation, but does not lead directly to estimates of risk at environmental exposure levels. Since an added risk of 50% will not always be achieved at the MTD, estimation of the TD_{50} may also require extrapolation outside the experimental dose range, albeit to a lesser degree than with q_1^*. Of 217 bioassays considered by Krewski et al. (1990b), for example, 65 of the TD_{50} values exceeded the MDT (cf. Munro, 1990). The need to extrapolate above the experimental dose range can be reduced by the use of a lower quantile of the dose-response curve, such as the TD_{25} employed by Allen et al. (1988a). (Note that the TD_{25} will *not* generally be equal to one-half of the TD_{50} in the presence of curvilinear dose response.)

Arguments in favor of the use of an even lower quantile of the dose response curve can be made. Crump (1984b) introduced the notion of a benchmmark dose for toxicological risk assessment, which corresponds to a quantile such as the TD_{10}. This benchmark dose is not strongly dependent on the dose-response model used to describe the data (Krewski et al., 1990a), and will likely lead to rankings similar to the TD_{50} or TD_{25}. Cogliano (1986) has recently shown that the TD_{10} is highly correlated with q_1^*; the TD_{10} could then be used as a starting point for linear extrapolation to lower doses, thereby providing a single index for potency ranking and low dose risk assessment. Other investigators have previously proposed linear extrapolation from the TD_{01} for low dose risk estimation (Mantel & Bryan, 1961; Van Ryzin, 1980; Farmer et al., 1982; Gaylor, 1983).

Another approach to estimating low dose risks is the model free extrapolation (MFX) method proposed by Krewski et al. (1991a). This procedure assumes only that the dose response curve is linear at low doses, and is based on a series of secant approximations to the slope of the dose response curve obtained by linear interpolation between points in the low dose region and controls. Upper confidence limits on the slope of the dose-response curve based on MFX are generally close to the values of q_1^* obtained from the LMS model. If the dose response curve is actually sublinear at low doses, the MFX method still provides an upper bound on low dose risks.

In practice, estimation of measures of carcinogenic potency such as the TD_{50} is not as straightforward as might appear from the preceding discussion. Ideally, estimation of the TD_{50} should take into account both intercurrent mortality in long-term animal studies and, when available, cause of death information (Finkelstein & Ryan, 1987; Finkelstein, 1991). Sawyer et al. (1984) propose methods for adjusting for intercurrent mortality with rapidly lethal tumors; Portier & Hoel (1987) show that estimates of the TD_{50} may be biased when the assumption of rapid tumor lethality is not satisfied. Dewanji et al. (1992) proposed a Weibull model that can be used for this purpose, provided that the survival times of individual animals are available for analysis. Bailar & Portier (1992) also use a Weibull model in estimating carcinogenic potency.

2.2 Carcinogenic Potency Database (CPDB)

Gold et al. (1984, 1986a, 1987, 1990) have tabulated the TD_{50} values for a large number of chemicals which have induced tumors in laboratory animals in their Carcinogenic Potency Database (CPDB). The TD_{50} values were calculated using statistical methods developed by Sawyer et al. (1984) and Peto et al. (1984) using a one-stage model. All TD_{50} values are expressed in common units of mg/kg body weight/day, adjusted to a standard two year rodent lifetime, and corrected for intercurrent mortality whenever individual animal data was available (Gold et al.,

1986b). When the level of exposure was not constant throughout the study period, a time-weighted average daily dose was used to determine the TD_{50}. Although more precise methods of estimating carcinogenic potency with time-dependent exposure patterns are available (Murdoch & Krewski, 1988), this is not critical for our present purposes (cf. Murdoch et al., 1992).

The CPDB includes data on over 3700 experiments on 975 different chemicals conducted under the National Cancer Institute/National Toxicology Program and by other investigators who have reported their results in the scientific literature (Gold et al., 1989). For each chemical, the CPDB may include studies done on different sexes, species and strains; by various routes of exposure; or under other experimental conditions. For each experiment, the doses and crude tumor response rates for each lesion demonstrating evidence of a dose-related effect are provided, thereby affording the opportunity for secondary analyses of the experimental results.

2.3 Variation in Carcinogen Potency

The CPDB includes data on potent chemical carcinogens such as 2,3,7,8-tetraclorodibenzo-p-dioxin (TCDD), as well as less potent compounds such as DDT. Gold et al. (1984) noted that the TD_{50} value in the CPDB for chemicals inducing tumors in rats varied by seven orders of magnitude or 10 million fold.

In studying the distribution of carcinogenic potency, Rulis (1986) noted that the TD_{50} values for 343 rodent carcinogens selected from the CPDB were roughly lognormally distributed. (In cases where more than one experiment was done on the same chemical or where more than one lesion was dose-related in a single study, the minimum TD_{50} value was used in this analysis.) Similar distributions have been observed using other subsets of the CPDB (Krewski et al., 1990b). For example, consider the distribution of TD_{50} values shown in Figure 1a for 191 of the 217 compounds considered previously by Krewski et al. (1990b). These compounds were selected to satisfy a number of criteria, including the requirements that the experiment have at least two doses in addition to unexposed controls and demonstrate clear evidence of carcinogenicity. The 26 experiments omitted from the current analysis included only one

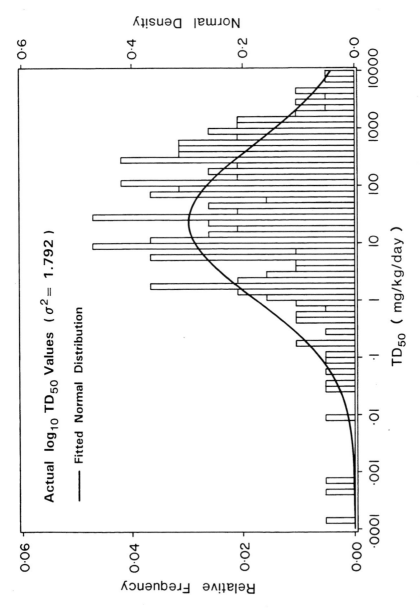

FIGURE 1a Variation in carcinogenic potency of 191 chemical carcinogens selected from the CPDB.

nonzero dose level, which precluded the use of the Weibull model to estimate the TD_{50} (see annex A for a detailed discussion of this issue). The median TD_{50} based on the fitted lognormal distribution was approximately 29 mg/kg/day, with 10th and 90th percentiles of 0.5 and 896 mg/kg/day, respectively. Although the inter-decile range is limited to a range of potencies of about 2,000-fold, the observed potencies again vary by more than eight orders of magnitude due in large part to the very low TD_{50} value for TCDD.

Rulis (1986) suggested that this distribution of carcinogenic potency could be used to establish a level of exposure below which regulatory attention would not be required. Such a threshold of regulation would be established on the basis of a lower quantile of the distribution of TD_{50}s, on the assumption that a new untested chemical would be unlikely to be more potent than most known animal carcinogens (Munro, 1990). This concept has also been considered by Zeise et al. (1984).

The distribution of TD_{50}s in Figure 1a is subject to overdispersion, since each individual TD_{50} is subject to experimental error. The distribution of true TD_{50}s may be determined using the shrinkage estimators described in annex B. This technique adjusts for overdispersion by "shrinking" each estimated TD_{50} towards the mean TD_{50} on a logarithmic scale, using a shrinkage factor determined by the relative magnitude of the variation within and between experiments (see annex B for details). Due to the large variation in TD_{50}s noted in different studies, and the comparatively small standard error for individual TD_{50}s, application of the shrinkage estimator in this case reduces the variance of the logarithm (base 10) of the TD_{50} from $\sigma^2 = 2.2$ to $\sigma^2 = 1.8$ (Figure 1b).

2.4 Classification of Carcinogens

Based on an evaluation of 237 chemical carcinogens tested in theU.S. National Toxicology Program, Rosenkranz & Ennever (1990) showed that carcinogens that are active at multiple sites in more than one species tend to be more potent than carcinogens that affect a single species or a single tissue. McGregor (1992) recently examined the characteristics of chemical carcinogens in different categories used by the International Agency for Research on Cancer to classify the strength of evidence for carcinogenicity. Carcinogens in Group 1 (known human carcinogens) tended to be more potent in rodents than carcinogens in Group 2A (pro-

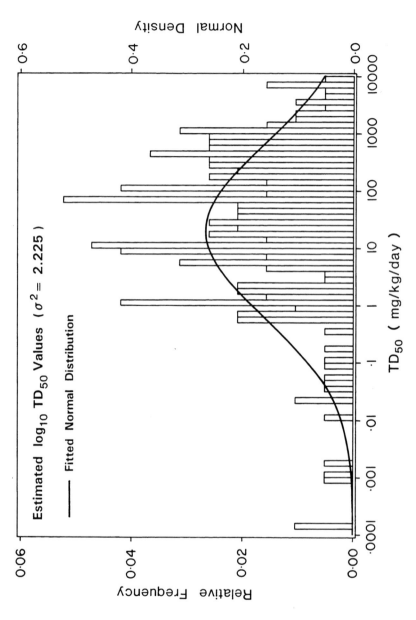

FIGURE 1b Variation in carcinogenic potency of 191 chemical carcinogens selected from the CPDB.

bable human carcinogen), Group 2B (possible human carcinogens), and Group 3 (unclassifiable with respect to human carcinogenicity).

Rosenkranz & Ennever (1990) also showed that genotoxic carcinogens that demonstrated mutagenic effects in the *Salmonella* assay were, on average, more potent than nongenotoxic carcinogens that tested negative in *Salmonella*. Human carcinogens also appear to be predominantly genotoxic (Shelby et al., 1988; Bartsch & Malaveille, 1989).

Based on an examination of the potency of carcinogens evaluated by the International Agency for Research on Cancer, McGregor (1992) concluded that there did not appear to be a strong association between carcinogenic potency in rodents and genotoxicity. It was noted that the most potent rodent carcinogen (TCDD) is apparently nongenotoxic, whereas one of the least potent rodent carcinogens (phenacetin) is mutagenic in the *Salmonella* assay.

3. Correlation Between TD_{50} and the MTD

3.1 Empirical Correlations

Several investigators have noted a marked correlation between carcinogenic potency and the MDT (Bernstein et al., 1985; Gaylor, 1989; Krewski et al., 1989; Reith & Starr, 1989a). To demonstrate this relationship, we reanalyzed data in the CPDB on the 191 chemical carcinogens discussed in section 2.3. Following Gold et al. (1984), we first used the one-stage model to estimate the TD_{50} for each carcinogen based on the crude proportion of animals developing tumors at each dose (Figure 2a). To allow for curvilinear dose-response, the TD_{50} was also estimated using both the multistage and Weibull models (Figures 2b and 2c respectively). These estimates of carcinogenic potency values are all adjusted to a standard two-year rodent lifetime as described in annex C. In each case, there is a strong positive association between the TD_{50} and the MDT, indicating that the most potent carcinogens in the database are those with the smallest MDTs.

The Pearson correlation coefficients between $\log_{10}(TD_{50})$ and $\log_{10}(MDT)$ are 0.924, 0.952 or 0.821, depending on whether the one-stage, multistage or Weibull model is used to estimate the TD_{50}. Note that the multistage model, which provides for curvature, yields a higher

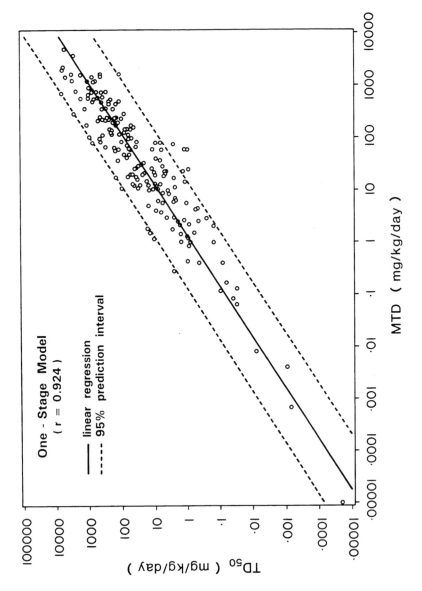

FIGURE 2a Association between carcinogenic potency and maximum tolerated dose.

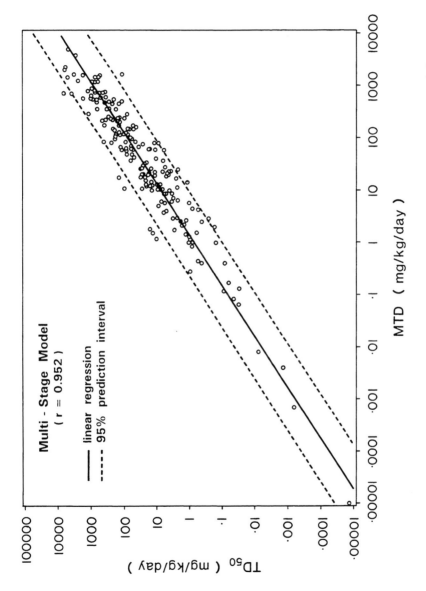

FIGURE 2b   Association between carcinogenic potency and maximum tolerated dose.

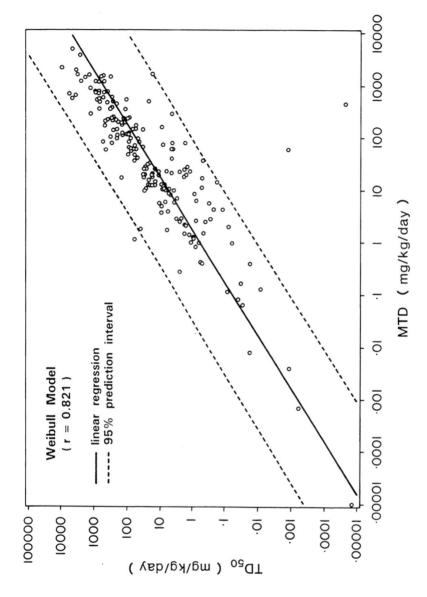

FIGURE 2c   Association between carcinogenic potency and maximum tolerated dose.

correlation than the essentially linear one-stage model. The lowest correlation arises from the Weibull model. This is because, unlike the one-stage or multistage models, the Weibull model allows for supra-linearity at low and moderate doses, and thus admits a greater range of TD_{50} values.

Crouch et al. (1987) commented on the absence of observations in the upper left and lower right triangular regions in scattergrams similar to those shown here in Figure 2. The absence of points in the upper left hand region is due to the lower limit on the number of tumors observed in the exposed groups in order to demonstrate a statistically significant increase in tumor occurrence. This implies that highly toxic chemicals of weak carcinogenic potency would likely go undetected in a standard bioassay, since such agents would not yield a measureable excess of tumors at the MTD. Crouch et al. (1987) attribute the absence of points in the lower right hand region to a lack of chemicals with extremely high potency relative to their MTDs. Reith & Starr (1989a) dispute this latter conclusion on the basis that experimental design constraints preclude the observation of potencies much larger than those shown in Figure 2. (This point is explored in greater detail in section 3.2 below.) Whether or not such "supercarcinogens" exist has been recently debated by the National Research Council (1992).

3.2   Range of Possible TD_{50} Values

Bernstein et al. (1985) noted that TD_{50} values calculated from bioassay data vary within a limited range about the MDT as a function of the observed tumor response. To illustrate, suppose that the probability $P(d)$ of a tumor occurring at dose d follows the one-stage model in (2.2), and the background tumor rate $P(0) = 1 - e^{-\alpha}$ is known to be 0.10. Suppose further that 50 animals are exposed to a single dose $D = MTD$ and that x of these animals develop the tumor of interest. Solving the equation

$$\frac{x}{50} = 1 - 0.9e^{-\beta D} \tag{3.1}$$

leads to the estimate

$$\hat{\beta} = -\frac{1}{D} \log_e\left(\frac{1-\frac{x}{50}}{0.9}\right) \qquad (3.2)$$

of β. We will consider tumor counts in the range $10 \leq x \leq 49$, the lower limit being the smallest value of x which is significantly greater (p < 0.05) than the assumed background incidences of 10%, and the upper limit being the largest value of x which leads to a meaningful estimate of β. (Since 100% tumor incidence is rarely observed, this truncated upper limit has little practical significance.) Under these conditions, we have $0.118/D \leq \beta \leq 3.807/D$, so that β could vary by about 32-fold. It follows from (2.3) that the corresponding estimate of carcinogenic potency varies by the same amount. Bernstein et al. (1985) assert that this result also tends to hold for more general experimental designs involving two or three exposed groups in addition to an unexposed control. It follows that since cancer potency values are constrained to lie within a narrow range determined by the MTD, the wide variation in MTDs for chemical carcinogens necessarily induces a high correlation between cancer potency and the MTD.

3.3 Analytical Correlations

The correlation between the MTD and the TD_{50} may be also be established using analytical arguments, details of which are provided in annex D. Suppose first that the MTD for a population of carcinogens follows a log-normal distribution as suggested by Bernstein et al. (1985). This assumption is supported by the approximate normality of the $\log_{10}TD_{50}$ values for the 191 chemicals discussed in section 3.1 (see Figure 3). (The normality assumption is actually not essential here since the correlation depends only on the variance of this distribution.) Suppose further that the TD_{50} is uniformly distributed about the MTD within the limits calculated the methods of by Bernstein et al. (1985). To the extent that these limits are greater than would be observed in practice, this assumption would tend to reduce the value of the correlation coefficient to be calculated by this argument. Under these assumptions, the correlation between $\log_{10}(TD_{50})$ and $\log_{10}(MTD)$ is 0.965 for a sample of size n = 50 animals in the exposed group.

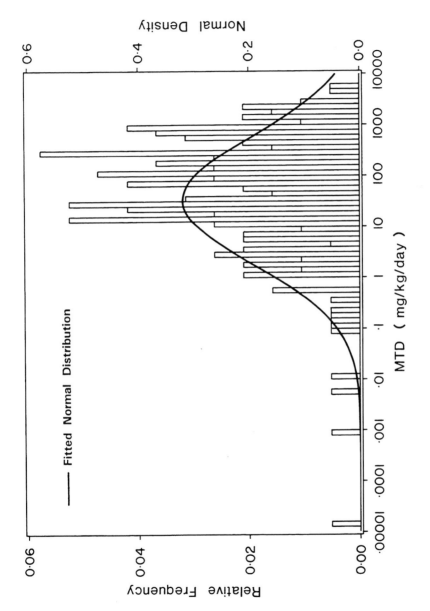

FIGURE 3 Distribution of MTDs for 191 chemical carcinogens selected from the CPDB.

To explore the extent to which this result is influenced by sample size, similar calculations were performed for a series of experiments with sample sizes ranging from n=50 to n=1000 animals per group (Table 1). These results indicate that the correlation between the $\log_{10}(TD_{50})$ and the $\log_{10}(MTD)$ remains high, even at the larger sample sizes for which the allowable range of potency values becomes much wider. Even in the limiting case of n = ∞, we have ρ = 0.944 (see annex D).

3.4 Model Dependency

Bernstein et al. (1985), Crouch et al. (1987), and Reith and Starr (1989ab) all used a one-stage model to characterize the carcinogenic potency. The one-stage model does not accommodate the majority of dose-response curves which exhibit curvilinearity. Kodell et al. (1990) argue that this limits the range of estimates of potency, and so artificially increases the correlation between the estimates of potency and the

TABLE 1   Correlation Between Carcinogenic Potency and the Maximum Tolerated Dose as a Function of Sample Size[a]

Sample Size n	Range of Experimental Outcomes, x/n		Range of Potency Estimates (upper limit ÷ lower limit)	Correlation[c]
	Minimum	Maximum		
50	0.200	0.98	32	0.965
100	0.150	0.99	79	0.957
500	0.122	0.998	247	0.950
1000	0.116	0.999	349	0.949
∞[b]	0.1	1.0	∞	0.944

[a] Based on a one-stage model and the assumptions in annex D.
[b] Limiting case as n → ∞.
[c] $\rho = \mathrm{Corr}(\log TD_{50}, \log_{10} MTD)$

MTD. Under the one-stage model, the potency β is related to the dose $D = MTD$ and the added risk $R(D)$ by

$$\log_e\beta = \log_e(1/D) + \log_e[-\log_e(1 - R(D))] . \qquad (3.3)$$

For a population of chemicals, this relationship provides a linear regression of $\log_e\beta$ versus $\log_e(1/D)$ with a slope of unity. The error term, $\log_e[-\log_e(1 - R(D))]$, expresses the variation in $R(D)$. Since the extra risk at the MTD is likely to fall in the range of 0.10 to 0.98, the variation about 1/MTD is limited to a range of approximately $\log_e[-\log_e(1 - 0.10)] = -2.25$ to $\log_e[-\log_e(1 - 0.98)] = 1.36$. This corresponds to the approximate 30-fold range noted by Bernstein et al. (1985). Using (2.3), the relationship in (3.3) may be re-expressed as

$$\log_e TD_{50} = \log_e D + \log_e[\log_e(1/2)/\log_e(1-R(D))] , \qquad (3.4)$$

so that the TD_{50} is restricted to this same range.

Kodell et al. (1990) suggest relaxing the linear restraints of the one-stage model and using the Weibull model in (2.5) to accommodate curvature. The Weibull model includes the one-stage model as a special case when $k = 1$. The TD_{50} and MTD are related by

$$\log_e(TD_{50}) = \log_e(D) + k^{-1} \log_e[\log_e(1/2)/\log_e (1 - R(d))], \qquad (3.5)$$

where k varies from chemical to chemical to accommodate either convexity (upward curvature, $k > 1$) or concavity (downward curvature, $k < 1$) in dose-response. This permits additional variation and reduces the correlation between $\log_e(TD_{50})$ and $\log_e$ (MTD) obtained with k fixed at unity. Bailar et al. (1988) demonstrate that an appreciable portion of the National Cancer Institute/National Toxicology Program bioassays exhibit downward curvature ($k < 1$) (cf. Williams & Portier, 1992). Clearly, the one-stage model limits the values of potency estimates to a rather narrow range determined by the MTD, which contributes to the observed correlation between $\log_e(TD_{50})$ and $\log_e$ (MTD). We expect that the MTD will still be highly correlated with the TD_{50} derived from a Weibull model, although the degree of correlation will be somewhat less than is observed with the one-stage model.

This expectation is confirmed by the correlation coefficients between the $\log_{10}(TD_{50})$ and $\log_{10}(MDT)$ reported in Figure 2 for the multistage and Weibull models. As further confirmation of this, we used the analytical approach in annex D to determine the correlation as a function of the shape parameter k in the Weibull model for a spontaneous tumor response rates of $P(0) = 0.10$ (Table 2). These results indicate that the correlation remains high for all values of k, increasing from about 0.9 for k near zero to almost 1 for large values of k.

3.5 Genotoxic vs. Nongenotoxic Carcinogens

Goodman & Wilson (1992) compared the dependence of the TD_{50} on the MTD for 217 genotoxic and nongenotoxic chemicals subjected to rodent bioassay within the U.S. National Toxicology Program. In this study, segregation of genotoxic and nongenotoxic carcinogens was done primarily on the basis of structural alerts and mutagenicity in *Salmonella* as described by Ashby & Tennant (1988). This analysis demonstrated that the TD_{50} for both genotoxic and nongenotoxic rodent carcinogens was highly correlated with the MTD. The authors found that the vari-

TABLE 2 Correlation[a] Between Carcinogenic Potency and the Maximum Tolerated Dose as a Function of the Weibull Shape Parameter k

Weibull Shape Parameter k	$\rho = $ Corr $(\log_{10}TD_{50}, \log_{10}MTD)$
0.0^b	0.944
0.5	0.946
1.0	0.965
3.0	0.994
5.0	0.998
∞^b	1.000

[a] Based on assumptions in annex D.
[b] Limiting cases as $k \rightarrow 0, \infty$.

ability about the fitted regression line

$$\log_e(1/TD_{50}) = a + b \log_e(1/MTD) \qquad (3.6)$$

was somewhat greater for genotoxic carcinogens than for nongenotoxic carcinogens, implying that this relationship is weaker for genotoxic carcinogens as compared to nongenotoxic carcinogens. The authors suggested that this is consistent with the hypothesis that carcinogenic effects observed at the MTD are mediated to a certain extent by toxicity, and that the larger variability exhibited by genotoxic carcinogens is because of their ability to induce carcinogenic effects by direct damage to genetic material. This does not imply that toxicity does not play a role in the induction of tumors by mutagenic chemicals; rather it is the inability of nongenotoxic agents to interact directly with DNA that leads to this difference.

Goodman & Wilson (1992) also examined a second set of 245 compounds which had tested positive in various *Salmonella* strains (cf. Zeiger et al., 1988). Since no significant differences in the variablity of fitted regressions lines were noted within three categories of mutagenic potency, it did not appear possible to further characterize the variability in the potency of genotoxic carcinogens relative to the MTD on the basis of genotoxic potency.

4.   Prediction of the TD_{50}

4.1   Predictions Based on the MDT

The high correlation between the TD_{50} and the MTD demonstrated in section 3 suggests that the TD_{50} may be predicted from the MTD. To explore this possibility, we fit the linear regression model

$$\log_{10}(TD_{50}) = a + b \log_{10}(MTD) + e \qquad (4.1)$$

to data on the 191 chemical carcinogens considered previously in section 2.3. Here, e represents a random error term which is assumed to be normally distributed with mean zero and variance σ^2, and a and b are parameters that can be estimated using ordinary least squares. Separate regressions were performed using the one-stage, multistage, and Weibull models to determine the TD_{50}.

As indicated in Table 3, the estimate of the slope is near unity for all three models. The error variance σ^2 is greatest for the Weibull model, and least for the one-stage model. An approximate 95% prediction interval for an individual $\log_{10}(TD_{50})$ at a given value of the MTD is given by $a + b\log_{10}(MTD) + 2\sigma$. Since $b \approx 1$, this corresponds to the interval $10^{\pm 2\sigma}$ x MTD for the TD_{50}. For the one-stage model, for example, the 95% prediction interval encompasses a range of 7 x 7 = 49 fold about the MDT, comparable to the 32-fold range derived by Bernstein et al. (1985).

4.2   Predictions Based on Mutagenicity and Acute Toxicity

The preceding results indicate that the MTD, which is essentially a measure of chronic toxicity, is a fair predictor of the TD_{50}. Determination of the MTD is normally based on the results of subchronic toxicity tests lasting about three months. This observation raises the question

TABLE 3   Regression of Carcinogenic Potency on the Maximum Tolerated Dose

Regression Parameter	Model		
	One-Stage	Multistage	Weibull
Intercept ± SE	- 0.07 ± 0.05	- 0.10 ± 0.04	0.15 ± 0.06
Slope ± SE	1.04 ± 0.02	1.04 ± 0.02	0.99 ± 0.03
Correlation	0.952	0.964	0.903
Root Mean Square (σ)	0.423	0.362	0.592
Factor $10^{2\sigma}$ for 95% Prediction Interval[b]	7.0	5.3	15.3

[a] Based on simple linear regression of $\log TD_{50}$ on log MDT
[b] Upper limit is $10^{2\sigma}$ x MDT; lower limit is $10^{-2\sigma}$ x MDT.

as to whether there exist other variables that are highly correlated with cancer potency that may be considered as possible predictors of the TD_{50}, and which may be determined with less effort than through a subchronic study.

The two-stage initiation-promotion-progression model of carcinogenesis described by Moolgavkar & Luebeck (1990) is based on the concept that mutation and cell proliferation are the two most important determinants of neoplastic change. Meselson & Russell (1977) reported a near-perfect linear relationship between the mutagenic potency of 14 chemical carcinogens, as measured by the dose inducing a mutation rate of 100 revertant colonies in the Ames *Salmonella*/microsome assay, and carcinogenic potency, as measured by the TD_{50}. McCann et al (1988) found a correlation of r = 0.41 between the mutagenic and carcinogenic potencies of 80 chemicals drawn from both the general literature and the U.S. National Toxicology Program. More recently, Piegorsch & Hoel (1988) examined the correlation between mutagenic and carcinogenic potency using 97 chemicals tested in the U.S. National Toxicology Program. In this analysis, mutagenic potency was measured by the slope of the initial linear portion of the dose response curve (cf. Krewski et al., 1992, 1993). Although a significant positive correlation of r=0.48 between mutagenic and carcinogenic potency was apparent, the overall scatter was considered to be to sufficiently great to preclude the use of Ames test data as a quantitative predictor of carcinogenic potency. Parodi et al. (1990) reviewed studies of the correlation between mutagenic and carcinogenic potency conducted between 1976 and 1988. In addition to using the *Salmonella* assay to measure mutagenic potency, mutation in L5178Y mouse lymphoma cells, *in vivo* DNA adducts in rodent liver, and *in vitro* DNA repair in rodent liver were also considered. This investigation suggests that the correlation between carcinogenic potency and mutagenic potency based on each of these short-term tests is moderate, with correlation coefficients in the neighborhood of r=0.4.

The relationship between acute toxicity, which may provide some indication of the ability of a chemical to induce cellular proliferation, and carcinogenic potency has been the subject of several investigations. Parodi et al. (1982a) found a significant correlation (r = 0.49) between carcinogenic potency and acute toxicity. Parodi et al. (1990) suggested that this association may be due in large part to the fact that acute and chronic toxicity are correlated, with chronic toxicity (as measured by the

MTD) in turn being highly correlated with carcinogenic potency.

Zeise et al. (1986) reported a high correlation between acute toxicity as measured by the LD_{50}, and carcinogenic potency, as measured by the parameter β in the one-hit model. Although correlation coefficients as high as $r = -0.93$ were found with both variables expressed on a logarithmic scale, some exceptions were noted. For example, the carcinogenic potency of 7,12-dimethyl benz(a)anthracene (DMBA) is about 5,000 fold greater than would be predicted on the basis of its LD_{50}. Nonetheless, Zeise et al. (1984) suggested that this relationship between acute toxicity and carcinogenicity could be used to substantially narrow the uncertainty in the TD_{50} values of untested carcinogens, which, as shown previously, vary over some seven orders of magnitude. More recently, Metzger et al. (1987) reported somewhat lower correlations ($r = 0.6$) between the LD_{50} and TD_{50} for 264 carcinogens selected from the CPDB, with an average TD_{50}/LD_{50} ratio of 0.06.

Travis et al. (1990a) argued that since both mutation and cell proliferation are important determinants of carcinogenesis, attempts to correlate carcinogenic potency with mutagenicity or acute toxicity alone are inadequate. Thus, Travis et al. (1990ab, 1991) investigated the correlation between the TD_{50} and composite indices based on mutation, toxicity, reproductive anomalies, and tumorigenicity data derived from the Registry of Toxic Effects of Chemical Substances (RTECS) (Sweet, 1987). In general, this analysis confirmed the previously reported correlation of $r = 0.4$ between mutagenic potency in the Ames assay and the minimum TD_{50} observed in rodent carcinogenicity studies (McCann et al., 1988; Piegorsch & Hoel, 1988); the correlation of $r = 0.7$ between LD_{50} and TD_{50} was also somewhat greater than that reported by Metzger et al. (1987).

In addition to confirming previous findings, Travis et al. (1990ab, 1991) investigated the correlation between composite predictors of carcinogenic potency based on results from two or more of 870 different assays for mutagenicity or toxicity, including 20 assays for mutation reported in RTECS. For each assay, a relative potency index was established in terms of weighted average of the potency relative to 20 reference compounds; a geometric mean of all available assays was then used to obtain an overall predictor of carcinogenicity. For some chemicals, the relative potency values based on different assays varied by as much as five orders of magnitude.

This series of analyses provided several interesting results. First, while the use of two or more assays for mutation increases the correlation with the TD_{50} beyond that obtained using the Ames test alone, the increase was not statistically significant. The use of mutation and acute toxicity data combined did however yield a significantly higher correlation ($0.76 \leq r \leq 0.85$, depending on the chemicals selected) than was obtained with the use of mutation or acute toxicity data alone. When the analysis was restricted to carcinogens affecting specific target organs (lung or liver), correlation coefficients in the neighborhood of $r = 0.9$ were obtained. Using all of the RTECS assays, the correlation of the composite relative potency index with the minimum TD_{50} across sites was $r = 0.80$, 0.87 or 0.79, depending on whether data for rats, mice, or the most sensitive species was used. Although this last index included any data on tumorigenicity available in RTECS, Travis et al. (1990a) noted that exclusion of the tumor data from the index did not appreciably alter the results obtained.

Recently, Goodman & Wilson (1992) calculated the correlation between the TD_{50} and LD_{50} for 217 chemicals that they classified as being either genotoxic or nongenotoxic. The correlation coefficent for genotoxic chemicals was approximately $r = 0.4$ regardless of whether rats or mice were used, whereas the correlation coefficient for nongenotoxic chemicals was approximately $r = 0.7$.

McGregor (1992) calculated the correlation between the TD_{50} and LD_{50} for different classes of carcinogens considered by the International Agency for Research on Cancer. The highest correlations were observed in Group 1 (known human carcinogens) with $r = 0.72$ for mice and $r = 0.91$ for rats, based on samples of size 9 and 8 respectively.

5. Low Dose Risk Assessment

5.1 Correlation Between Upper Bounds On the Low Dose Slope and MTD

Krewski et al. (1989) noted that the values of q_1^* derived from the linearized multi-stage model fitted to 263 data sets were highly correlated on a logarithmic scale with the MDTs in those experiments. As with the TD_{50}, this association between q_1^* and the MDT occurs as a result of the limited range of values q_1^* can assume once the MTD is established.

This behavior is readily illustrated using model-free upper bounds on low dose risk proposed by Krewski et al. (1991a). The current NTP carcinogenesis screening bioassay generally consists of groups of 50 animals at doses of 0, D/4, D/2 and D = MTD. Using the model-free extrapolation (MFX) procedure with this design, the lowest estimate of potency would occur at the highest possible dose not exhibiting a statistically significant increase in tumor incidence. For carcinogens, there would be a statistically significant increase in the tumor incidence at least at the MTD. Hence, the lowest estimate of potency occurs when there are no tumors at the MTD/2. In this case, the MFX would yield an upper confidence limit on the low dose slope of approximately 0.09/(MTD/2) = 0.18/MTD. The maximum estimate of the low-dose slope would occur if the upper confidence limit on the incidence were 1.0 at the lowest dose, i.e., 1.0/(MTD/4) = 4/MTD. Using the MFX procedure, this design can only accommodate a 4/0.18 = 22-fold range in carcinogenic potency estimates. This would be reduced to an 11-fold range if the lowest dose (MTD/4) were omitted.

The strong negative association between the MDT and linearized upper bounds on the slope of the dose response curve in the low dose region is demonstrated empirically in Figure 4 using the 191 carcinogens considered previously in section 3.1. Upper bounds based on both the multistage model and model-free extrapolation are highly correlated with the MDT, with Pearson correlation coefficients of -0.941 and -0.960 respectively.

5.2   Correlation Between q_1^* and the TD_{50}

The fact that both the TD_{50} and q_1^* are correlated with the MTD implies a correlation between the TD_{50} and q_1^*. Krewski et al. (1989) provided empirical confirmation of this. An association between the TD_{50} and linearized estimates of low-dose cancer risks has been previously assumed by other investigators. Rulis (1986) used simple linear extrapolation from TD_{50} values in the CPDB to estimate the 10^{-6} risk-specific doses (RSDs) for 343 chemical carcinogens. (The RSD is the dose associated with a specified increase in risk.) Similarly, by defining

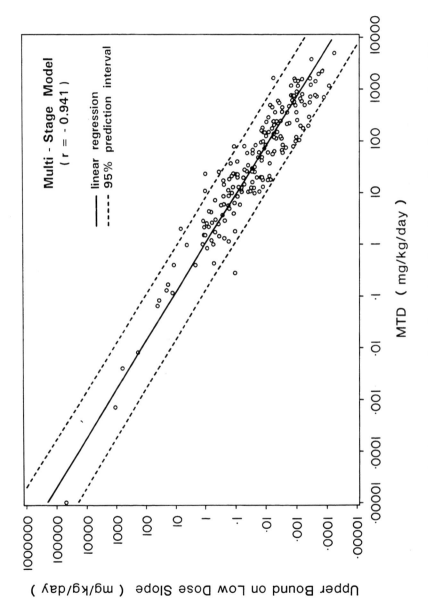

FIGURE 4a Association between upper bounds on low dose slope and maximum tolerated dose.

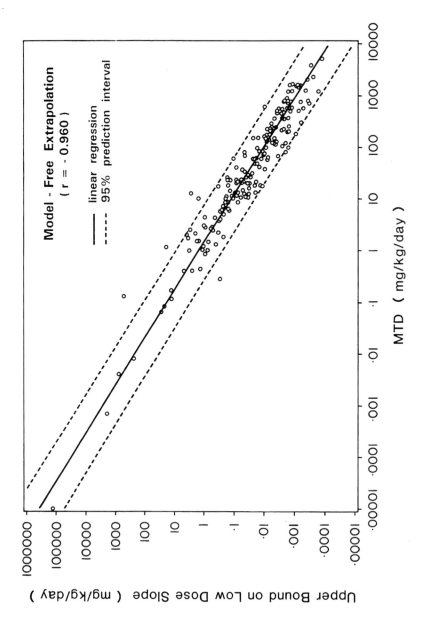

FIGURE 4b Association between upper bounds on low dose slope and maximum tolerated dose.

their HERP (Human Exposure/Rodent Potency) index as the fraction of the TD_{50} accounted for by human exposure, Ames et al. (1987) implicitly assume a linear dose response relationship below the TD_{50}. Wartenburg and Gallo (1990) have objected to the latter application of the TD_{50} on the grounds that many dose response relationships are highly nonlinear. In practice, however, linear extrapolation from the TD_{50} will often approximate q_1^* in experiments employing only two or three doses because of the limited opportunity to observe curvature (Krewski, 1990). Although the HERP index appears to be based on the tacit assumption of a linear dose response, Gold & Ames (1990) and Gold et al. (1992) emphasize that the index is intended for priority ranking rather than quantitative risk assessment.

5.3. Preliminary Estimate of Risk

The fact that q_1^* is highly correlated with the MDT suggests that preliminary estimates of cancer risk may be based on the MTD. Gaylor (1989) exploited this correlation to estimate the 10^{-6} RSD. Estimates of the RSD were obtained by the linear interpolation procedure given by Gaylor & Kodell (1980), as modified by Farmer et al. (1982), for 38 chemical carcinogens tested by oral administration in the U.S. National Toxicology Program. Estimates of the RSD were compared with the MTD for up to 69 tumor sites in both rats and mice, for a total of 138 cases. The ratio of the MTD to the RSD varied over a 184-fold range, which is considerably larger than the 32-fold range suggested by Bernstein et al. (1985) for the range of TD_{50} values relative to the MTD. The overall geometric mean of the ratio MDT/RSD was 3.8×10^5; only 3 of the 138 ratios were more than a factor of 10 from the mean. This suggests that a preliminary estimate of the RSD may be obtained by dividing the MTD by 380,000.

As in predicting the TD_{50} from the MDT (section 4.1), linear regression analysis may also be used to predict low dose slopes from the MDT. This may be illustrated using the 191 rodent carcinogens considered previously. The results of regressing the logarithms of the linearized upper bounds on low dose slope based on either the LMS model or MFX are given in Table 4. The estimated slope of the linear regression model is approximately -1 for both the LMS and MFX methods. The

TABLE 4   Regression of Upper Bounds on Low Dose Slopes on the
Maximum Tolerated Dose[a]

Regression Parameter	Extrapolation Method	
	Multistage Model	Model-Free Extrapolation
Intercept ± SE	0.01 ± 0.05	0.11 ± 0.04
Slope ± SE	-1.05 ± 0.03	-1.07 ± 0.02
Correlation	0.944	0.961
Root Mean Square (σ)	0.462	0.386
Factor $10^{2\sigma}$ 95% Prediction Interval[b]	8.4	5.9

[a] Based on simple linear regression of log slope on log MDT.
[b] Upper limit is $10^{2\sigma}$ x MDT; lower limit is $10^{-2\sigma}$ x MDT.

approximate 95% prediction intervals for the low dose slope encompass
a range of about 8 x 8 = 64 fold about the MDT with the LMS model,
and a range of about 36-fold for MFX. Given an upper bound on the
low dose slope β, the corresponding 10^{-6} RSD is simply $10^{-6}/\beta$.

6.   Interspecies Extrapolation

Since mammalian species share many common physiological characteristics it is expected that they may respond in a somewhat similar
manner to toxic substances. While many differences exist between
species (Oser, 1981), allometric relationships among physiological parameters have suggested different metrics for quantitative interspecies
extrapolation: heat loss, for example, appears to be proportional to the
surface area of mammals, whereas metabolism is related to body weight
to the 3/4 power (Schmidt-Nielsen, 1984). Such considerations have led
to suggestions that allometric equations of the form

$$P = a \times BW^b$$

may be used to relate carcinogenic potency (P) to body weight (BW). A value of b = 1 corresponds to interspecies extrapolation on the basis of body weight, whereas b = 2/3 corresponds to extrapolation roughly on the basis of body surface area. Travis & White (1988) suggest an intermediate value of b = 3/4 (cf. Watanabe et al., 1992), which corresponds roughly to extrapolation on the basis of metabolic rate (Schmidt-Nielsen, 1984).

6.1 Extrapolation from Rats to Mice

Quantitative interspecies extrapolation of measures of carcinogenic potency such as the TD_{50} may also be based on empirical observations of potency in the two target species, provided that the agents of interest are effective in both species. Crouch & Wilson (1979) demonstrated a positive correlation between rats and mice in carcinogenic potency expressed in terms of the slope coefficient β in the one-hit model in (2.2). Subsequently, Crouch (1983) suggested that interspecies extrapolation of carcinogenic potency would generally be accurate to within a factor of about 4.5.

Gaylor & Chen (1986) compared the relative carcinogenic potency of chemicals in rats, mice, and hamsters based upon the TD_{50}s given by Gold et al. (1984). Since current practice generally is to base risk estimates upon the data set producing the highest cancer risk, the minimum TD_{50} was selected for each chemical in each species for each route of administration. The largest subset of relative potencies was obtained for rats and mice for 190 chemicals administered in the diet. With dose expressed in terms of mg/kg body weight/day, the geometric mean of the ratio of the minimum TD_{50} value for rats relative to that for mice was 0.45. For dose expressed in terms of concentration (ppm) in the diet, however, the mean ratio was 1.3. Using either dose metric, the mean carcinogenic potency of these chemicals in rats and mice agree to within a factor of about two-fold. The ratio R of the TD_{50} values for rats and mice were approximately log-normally distributed, with $\log_{10}R$ exhibiting a standard deviation of 0.82; this corresponds to a multiplicative factor of $10^{0.82} \approx$ 7-fold. For 4 of the 190 chemicals, the ratios of the TD_{50}s for rats and mice differed by more than a factor of 100. Chen & Gaylor (1987) used results from the NCI/NTP Carcinogenesis Bioassay Program to compare cancer risk estimates for rats relative to those for mice for chemicals administered orally. The 10^{-6} RSD was calculated for rats and mice for those chemicals that showed a dose-response trend in the same sex at the same tissue/organ site in both species. In all, 69 comparisons of RSDs between rats and mice for 38 rodent carcinogens were made. The overall geometric mean of the RSD ratios for rats to mice was 1.27, with dose was measured in terms of concentration (ppm) in the diet. The logarithms of the ratios of RSDs were approximately normally distributed with a standard deviation of 0.79, corresponding to a multiplicative factor of approximately 6-fold. The RSD ratios varied from 1:51 to 49:1 for the 69 cases. Without the restriction of tumors at the same sex and site in both species, the geometric mean of the ratio of the minimum RSDs of rats to mice was 1.38 with a standard deviation of $\log_e$ ratios of 0.78. It appears that relative potencies for rats and mice are generally within a multiplicative factor of 100-fold for rodent carcinogens. However, McGregor (1992) recently noted that amides and halides tended to exhibit disparate TD_{50}s in rats and mice.

Bernstein et al. (1985) suggested that this apparent interspecies correlation in carcinogenic potency simply reflects the corresponding high correlation in MTDs for rats and mice. This provoked a debate as to the interpretation of these results on interspecies potency correlation (Crouch et al., 1987; Reith & Starr, 1989b).

Reith & Starr (1989b) obtained a correlation of r = 0.83 on a logarithmic scale between potency estimates for n = 83 chemicals selected from the CPDB identified as carcinogens in both rats and mice. (In this analysis, potency was defined as the slope β in (2.3), calculated using the TD_{50} values given in the CPDB.) They argued that the correlations arise from (i) the strong interspecies correlation between MTDs in chronic bioassays, (ii) the small numbers of animals used per dose group, and (iii) the narrow range of doses typically tested. Reith & Starr (1989b) further noted a high correlation for chemicals testing negative in both species (r = 0.85, n = 51), for chemicals testing positive in rats but negative in mice (r = 0.55, n = 15), and for chemicals testing negative in rats but positive in mice (r = 0.68, n = 25). Reith & Starr (1989b) recomputed these correlations after dividing each potency estimate by the MDT. The largest correlation ($r = 0.27$) was obtained for chemicals testing negative in both rats and mice; none of the four recomputed correlations was significantly greater than zero ($p > 0.05$).

To further illustrate the correlation in TD_{50} values for rats and mice, consider the data on 127 of the 492 rodent carcinogens discussed by Gold et al. (1989) which are carcinogenic in both species. These data demonstrate a high correlation between TD_{50} values for rats and mice (Figure 5), with a Pearson correlation of 0.808.

This observation may also be derived analytically (annex E). Let us assume that the MDTs for the rat and mouse carcinogens are both lognormally distributed, with $MTD_{rats} = cMTD_{mice}$. (Although Bernstein et al., 1985, estimate c to be 0.357, the correlation coefficient is independent of c.) Suppose further that the TD_{50}s for each species are uniformly distributed about the MTD within the 32-fold range considered by Bernstein et al. (1985), and that, given the MTDs for each species, the TD_{50}s for rats and mice are statistically independent. These assumptions lead to a correlation based on equation (E.5) in annex E of 0.943 for the TD_{50} values for rats and mice. The assumption of strict proportionality between MTD rats and MTD mice may be relaxed as discussed in annex E, leading to a reduced correlation of 0.763.

Shlyakhter et al. (1992) studied the correlation between carcinogenic potency and the MTD and the correlation between carcinogenic potencies in rats and mice by computer simulation based upon characteristics of NCI/NTP carcinogenicity tests. This investigation demonstrated that the observed correlation between carcinogenic potency and the MTD could, under certain conditions, produce a correlation which is purely artifactual. However, by comparison with actual bioassay data it was concluded that the observed correlation cannot be an artifact of constraints on the data and therefore must have some biological basis. This suggests that the observed correlation in carcinogenic potency between rats and mice cannot be attributed solely to bioassay design (particularly the MTD), so that the correlation is at least partly attributable to the biological similarity of rodent species.

Freedman et al. (1992) also argue that the correlation in carcinogenic potency between rats and mice is not entirely tautological. This analysis is based on a comparison of models for interspecies correlation that are either entirely artifactual (due to constraints imposed by the MTD) or which include a real component. The latter models lead to slightly higher correlations, suggesting that although empirically observed correlations such as that in Figure 5 are largely due to the correlation between the corresponding MTDs, at least part of this association is nonartifactual.

6.2 Extrapolation from Rodents to Humans

Allen et al. (1988ab) and Crump et al. (1989) compared the carcinogenic potency, as measured by the TD_{25}, for 23 chemcials for which suitable dose-response data were available from both human epidemiological studies and animal carcinogenesis bioassays. Several alternative methods of analyzing the animal bioassays were investigated, including the choice of the interspecies dose scaling factor, benign and malignant versus malignant tumors only, and separate versus average results across studies. Most of the methods yielded animal TD_{25}s, that were significantly correlated with human TD_{25}s, with rank correlation coefficients ranging up to 0.9. Although the correlation between potency rankings in animals and humans is high, the error associated with predictions of carcinogenic potency in humans based on animal data is substantial.

Chemotherapeutic agents represent another data base which may be useful in establishing animal-human correlation in carcinogenic potency. Kaldor et al. (1988) obtained estimates of the carcinogenic potency of 15 antineoplastic drugs that increase the risk of secondary tumors (acute non-lymphocytic leukemia). Two sets of TD_{50} values were obtained from the CPDB for 5 of these agents; the first set involved tumors of any type whereas the second set was restricted to tumors of the hematopoietic system. The potency rankings for 3 nitrogen mustard compounds (cyclophosphamide, chlorambucil, and melphalan) were similar in animals and humans. The most potent of the rat carcinogens, actinomycin D, did not cause leukemia in humans at the doses used, which are limited by its toxicity. The authors found these results encouraging in terms of using animal data to predict potency rankings, but cautioned against quantitative prediction of human carcinogenic potency on the basis of these data.

Goodman & Wilson (1991b) evaluated predictions of cancer risks based on potency values in the CPDB against epidemiological observa-

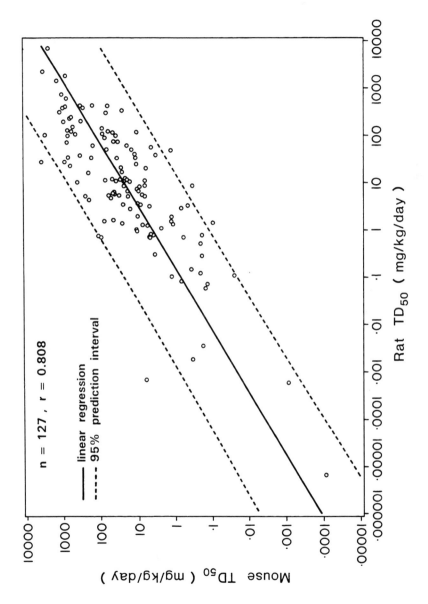

FIGURE 5   Association between carcinogenic potency in rats and mice.

tions on 22 of 29 chemicals considered previously by Ennever (1987) for which positive rodent bioassay data is available. These chemicals are of interest in that the epidemiological data did not provide unequivocal evidence of carcinogenic effects in humans: many were in fact in category 3 (i.e., not classifiable with respect to human carcinogenicity) within the classification scheme used by the International Agency for Research on Cancer (1987), and none were in category 1 (sufficient evidence of carcinogenicity in humans). Based on their re-analysis of this data, Goodman & Wilson (1991b) argued that the excess risks observed in epidemiological studies (which may or may not have been significantly elevated) were roughly consistent with predictions based on potency values in the CPDB. Goodman & Wilson (1991a) recently reviewed interspecies comparisons of carcinogenic potency and concluded that "there is a good correlation of the carcinogenic potencies between rats and mice, and the upper limits on potencies in humans are consistent with rodent potencies for those chemicals for which human exposure data are available".

7. Conclusions

The completion of a large number of laboratory studies of the carcinogenic potential of chemicals has afforded an opportunity to evaluate the variation in the potency of chemical carcinogens. The Carcinogenic Potency Database developed by Gold et al. (1984) provides a convenient summary not only of the data from nearly 4,000 individual experiments, but also of the potency of chemical carcinogens expressed in terms of the TD_{50}. The TD_{50}s in the CPDB indicate that carcinogenic potency may vary by nearly 10 million fold.

Several investigators have reported a strong correlation between the maximum dose tested (MDT) in carcinogen bioassay and the TD_{50}, which generally corresponds to the maximum tolerated dose (MTD). In particular, the estimate of the TD_{50} based on the one-hit model can be shown, using both theoretical and empirical arguments, to be restricted to lie within a factor of about 32-fold of the MTD. Empirical evidence indicates that measures of carcinogenic risk at low doses, such as the value of q_1^* in the linearized multistage model, are also correlated with the MTD, suggesting that preliminary estimates of low dose cancer risk may be based on an estimate of the MTD. Specifically, Gaylor (1989) has shown that dividing the MTD by a factor of 380,000 will approximate the 10^{-6} RSD obtained from bioassay data using the linearized multistage model.

Carcinogenic potency has also been shown to be somewhat correlated with both acute toxicity and mutagenicity, both of which are important factors in neoplastic change. In particular, target tissue toxicity may lead to proliferation of preneoplastic cells, and hence increase the pool of cells available for malignant transformation. Travis et al. (1991) have demonstrated a strong correlation between a composite index based on toxicity and mutagenicity and carcinogenic potency as measured by the TD_{50}. These results suggest that data on toxicity and mutagenicity may be combined to reduce the uncertainty in the carcinogenic potential of chemicals not yet subjected to long-term carcinogen bioassay.

The apparent correlation between acute toxicity and carcinogenicity does not imply a causal relationship between toxicity and carcinogenicity. The establishment of a causal relationship between toxicity and carcinogenicity presupposes a biological relationship between these two endpoints. In this regard, Hoel et al. (1988) noted little association between toxic tissue injury and neoplastic change in NTP studies. Clayson & Clegg (1991), however, discuss specific examples in which toxicity plays an important role in carcinogenesis. Parodi et al. (1982b) note that covalent binding with macromolecules, which can influence the mutagenic potency of chemicals, can also induce toxicity in some cases.

While these empirically derived correlations are of considerable interest, a clear interpretation of these findings in either biological or statistical terms remains to be accomplished. To be biologically meaningful, the rationale for such associations should be toxicologically plausible. While toxic, mutagenic, and carcinogenic effects do share certain characteristics in common, each of these processes is sufficiently complex to cast doubt on a causal relationship between simple measures of toxic and mutagenic potential and carcinogenic potency. Statistically, correlations between the MDT and the TD_{50} occur as a result of the narrow range of possible potency values within a single experiment in relation to the wide variation observed in the potency of chemical carcinogens. This has led to suggestions that the observed correlation between the MTD and the TD_{50} may simply be an artifact of the experimental designs currently used in carcinogen bioassay. In this regard, Reith & Starr

(1989a) concluded that "the chronic rodent bioassay, in and of itself, is altogether inadequate as a data source for estimating the risk to humans from exposure to carcinogenic agents".

In our view, correlations between the MTD and measures of cancer potency reflect the limited amount of information on cancer risks provided by carcinogen bioassay data. Once the MTD has been determined, TD_{50} and q_1^* values are somewhat insensitive to the experimental results and are constrained to lie within a narrow range, particularly when viewed in light of the eight order of magnitude variation in TD_{50} values for chemical carcinogens. This does not imply that estimates of carcinogenic potency based on bioassay data are not meaningful, but does demonstrate that both the TD_{50} and q_1^* represent relatively crude indicators of risk. At the same time, however, the value of q_1^* does represent the smallest possible linearized upper bound on low dose risk based on the multistage model which is consistent with the experimental data. The TD_{50}, moreover, represents a dose which has been shown, often without the need for extrapolation outside of the observable response range, to reduce the proportion of tumor-free animals by one-half.

Measures of carcinogenic potency such as the TD_{50} have also been shown to be highly correlated between different rodent species (rats and mice). Although this appears to offer support for quantitative interspecies extrapolation of cancer bioassay data, it is possible that this correlation may be largely due to the high correlation between the MTDs for different rodent species. Kaldor et al. (1988) have suggested that because of the relationship between animal $LD_{50}s$ and the doses of antineoplastic agents used in cancer chemotherapy, the apparent correlation in potency of these agents in animals and humans may be explained in part by toxicity considerations. Despite this correlation, the error associated with quantitative interspecies extrapolations of carcinogenic potency values can be 100-fold or greater.

Imperfect qualitative agreement between species also suggests the need for caution in quantitatively extrapolation between species (Freedman & Zeisel, 1988). Although all known human carcinogens are also carcinogenic in animals (Tomatis et al., 1989), concordance between rats and mice with chemicals tested in the U.S. National Toxicology Program is only about 74% (Haseman & Huff, 1987). Gold et al. (1989) subsequently reported on overall concordance between rats and mice of 76% for 392 chemicals in the CPDB. Piegorsch et al. (1992) note that species concordance depends on carcinogenic potency, and that for weak carcinogens, the maximum possible species concordance may be only about 80%. Lave et al. (1988) suggested that concordance between rats and mice may represent an upper bound on concordance between rodents and humans. Quantitative interspecies extrapolation of carcinogenic potency is therefore done under the presumtion that the agent in question will be effective in both species involved.

If progress in carcinogenic risk assessment based on bioassay data is to be made, it seems that additional information beyond that contained in traditional experiments is required. In particular, studies of the mechanisms of chemical carcinogenesis may provide new insights on the estimation of low dose risk (Moolgavkar & Luebeck, 1990). The relative importantance of mutation and cell proliferation in carcinogenesis particularly requires further discussion. Cohen & Ellwein (1990) show that proliferation of urinary bladder tissue is essential for the induction of bladder tumors with 2-acetylaminofluorene. Cunningham et al. (1991) recently demonstrated that 2,4-diaminotoluene (2,4-DAT) and 2,6-diaminotoluene (2,6-DAT) are equally mutagenic in *Salmonella*, yet only 2,4-DAT produces a sufficient increase in cell turnover in rat liver to lead to hepatocarcinogenesis. Ames & Gold (1990) conclude that "without studies of the mechanism of carcinogenesis, the fact that a chemical is a carcinogen at the MTD in rodents provides no information about low-dose risk to humans". Physiologically based pharmacokinetic models may afford an opportunity to increase the accuracy of risk estimates through improved tissue dosimetry (Krewski et al., 1991b); measurement of metabolic parameters in different species may also lead to improved interspecies extrapolation (Andersen et al., 1987). More sensitive indicators of effects at very low doses, such as markers of DNA damage suspected to play a role in neoplastic conversion (cf. Lutz, 1990), may also serve to provide improved estimates of risk in the future. All of these considerations suggest a more biologically based approach to cancer risk assessment is needed (Clayson, 1987).

8. Acknowledgements

We are grateful to Drs. David Clayson, Kenny Crump, Lois Gold, Lester Lave, Mary Paxton, and Marvin Schneiderman for helpful comments on this article. An earlier version of this paper was presented on September 6, 1990 in Washington, D.C. at the Workshop on Maximum Tolerated Dose: Implications for Risk Assessment sponsored by the U.S. National Academy of Sciences Committee on Risk Assessment Methodology, and discussed by Drs. Edmund Crouch and Lauren Zeise. A draft of this paper was also presented at the International Environmetrics Conference held in Como, Italy, from September 27-October 2, 1990.

9. References

Allen, B., Crump, K. & Shipp, A. (1988a). Carcinogenic potencies of chemicals in animals and humans. *Risk Analysis 8*, 531-544.

Allen, B., Crump, K. & Shipp, A. (1988b). Is it possible to predict the carcinogenic potency of a chemical in humans using animal data? (with discussion). In *Banbury Report 31: Carcinogen Risk Assessment: New Directions in the Qualitative and Quantitative Aspects*. Cold Spring Harbor Laboratory, Cold Spring Harbor, pp. 197-209.

Ames, B.N., Magaw, R. & Gold, L.S. (1987). Ranking possible carcinogenic hazards. *Science 236*, 271-280.

Ames, B.N. & Gold, L.S. (1990). Too many rodent carcinogens: mitogenesis increases mutagenesis. *Science 249*, 970-971.

Andersen, M.A., Clewell, H.J., Gargas, M.L., Smith, F.A. & Reitz, R.H. (1987). Physiologically based pharmacokinetics and the risk assessment process for vinyl chloride. *Toxicology and Applied Pharmacology 87*, 185-205.

Apostolou, A. (1990). Relevance of maximum tolerated dose to human carcinogenic risk. *Regulatory Toxicology and Pharmacology 11*, 68-80.

Armitage, P. (1985). Multistage models of carcinogenesis. *Environmental Health Perspectives 63*, 195-201.

Ashby, J. & Tenant, R.W. (1988). Chemical structure, Salmonella mutagenicity and extent of carcinogenicity as indicators of genotoxic carcinogenesis among 222 chemicals tested in rodents by the U.S. NCI/NTP. *Mutation Research 204*, 17-115.

Bailar, A.J. & Portier, C.J. (1992). An index of tumorigenic potency. *Biometrics*. In press.

Bailar, J.C., III, Crouch, A.E.C., Shaikh, R. & Spiegelman, D. (1988). One-hit models of carcinogenesis: conservative or not? *Risk Analysis* 8, 485-497.

Barr, J.T. (1985). The calculation and use of carcinogenic potency: a review. *Regulatory Toxicology and Pharmacology* 5, 432-459.

Bartsch, H. & Malaveille, C. (1989). Prevalence of genotoxic carcinogens among animal and human carcinogens in the IARC Monograph Series. *Cell Biology and Toxicology* 5, 115-127.

Bernstein, L., Gold, L.S., Ames, B.N., Pike, M.C. & Hoel, D.G. (1985). Some tautologous aspects of the comparison of carcinogenic potency in rats and mice. *Fundamental and Applied Toxicology* 5, 79-87.

Butterworth, B.E. (1990). Consideration of both genotoxic and nongenotoxic mechanisms in predicting carcinogen potential. *Mutation Research* 239, 117-132.

Carr, C.J. & Kolbye, A.C.(Jr.) (1991). A critique of the use of the maximum tolerated dose in bioassays to assess cancer risks from chemicals. *Regulatory Toxicology and Pharmacology* 14, 78-87.

Chen, J.J. & Gaylor, D.W. (1987). Carcinogenic risk assessment: comparison of of estimated safe doses for rats and mice. *Environmental Health Perspectives* 72, 305-309.

Clayson, D.B. (1987). The need for biological risk assessment in reaching decisions about carcinogens. *Mutation Research* 185,243-269.

Clayson, D.B. & Clegg, D.J. (1991). Classification of carcinogens: polemics, pedantics, or progress? *Regulatory Toxicology and Pharmacology* 14, 147-166.

Clayson, D.B., Iverson, F. & Mueller, R. (1992). An appreciation of the maximum tolerated dose: an inadequately precise point in the designing a carcinogesis bioassay. *Teratogenesis, Carcinogenesis, and Mutagenesis*. In press.

Cogliano, V.J. (1986). The U.S. EPA's methodology for adjusting the reportable quantities of potential carcinogens. *Proceedings of the 7th National Conference on Management of Uncontrollable Hazardous Wastes (Superfund '86)*. Hazardous Wastes Control Institute, Washington, DC, 182-185.

Cohen, S.M. & Ellwein, L.B. (1990). Cell proliferation in carcinogenesis. *Science* 249, 1007-1011.

Crouch, E.A.C. & Wilson, R. (1979). Interspecies comparison of carcinogenic potency. *Journal of Toxicology and Environmental Health* 5, 1095-1118.

Crouch, E.A.C. & Wilson, R. (1981). Regulation of carcinogens (with discussion). *Risk Analysis* 7, 47-66.

Crouch, E.A.C. (1983). Uncertainties in interspecies extrapolation of carcinogenicity. *Environmental Health Perspectives* 50, 321-327.

Crouch, E.A.C., Wilson, R. & Zeise, L. (1987). Tautology or not tautology? *Journal of Toxicology and Environmental Health* 20, 1-10.

Crump, K.S. (1984a). An improved procedure for low-dose carcinogenic risk assessment for animal data. *Journal of Environmental Pathology, Toxicology and Oncology* 5, 339-348.

Crump, K.S. (1984b). A new method for determining allowable daily intakes. *Fundamental and Applied Toxicology* 4, 854-871.

Crump, K., Allen, B. & Shipp, A. (1989). Choice of a dose measure for extrapolating carcinogenic risk from animals to humans: an empirical investigation of 23 chemicals (with discussion). *Health Physics* 57 (Supplement 1), 387-393.

Cunningham, M.L., Foley, J., Maronpot, R.R. & Matthews, H.B. (1991). Correlation of hepatocellular proliferation with hepatocarcinogenicity induced by the mutagenic noncarcinogen:carcinogen pair - 2,6- and 2,4-diaminotoluene. *Toxicology and Applied Pharmacology* 107, 562-567.

Dewanji, A., Krewski, D. & Goddard, M.J. (1992). A Weibull model for estimating tumorigenic potency. *Biometrics*. In press.

Ennever, F.K., Noonan, T.J. & Rosenkranz, H.S. (1987). The predictivity of animal bioassays and short-term genotoxicity tests for carcinogenicity and non-carcinogenicity to humans. *Mutagenesis* 2, 73-78.

Farmer, J.H., Kodell, R.L. & Gaylor, D.W. (1982). Estimation and extrapolation of tumor probabilities from a mouse bioassay with survival/sacrifice components. *Risk Analysis* 23, 27-34.

Finkelstein, D.M. & Ryan, L.M. (1987). Estimating carcinogenic potency from a rodent tumorigenicity experiment. *Applied Statistics* 36, 121-133.

Finkelstein, D.M. (1991). (1991). Modeling the effect of dose on the lifetime tumor rats from an animal carcinogenicity experiment. *Biometrics* 47, 669-680.

Freedman, D.A. & Zeisel, H. (1988). From mouse-to-man: the quantitative assessment of cancer risk. *Statistical Science* 3, 3-56.

Freedman, D.A., Gold, L.S. & Slone, T.H. (1992). How tautological are inter-species correlations of carcinogenic potencies? *Technical Report No. 334, Department of Statistics, University of California, Berkeley.*

Gaylor, D.W. (1983). The use of safety factors for controlling risk. *Journal of Toxicology and Environmental Health* 11, 329-336.

Gaylor, D.W. (1989). Preliminary estimates of the virtually safe dose for tumors obtained from the maximum tolerated dose. *Regulatory Toxicology Pharmacology* 9, 1-18.

Gaylor, D.W. and Kodell, R.L. (1980). Linear interpolation algorithm for low dose risk assessment of toxic substances. *Journal of Environmental Pathology and Toxicology* 4, 305-312.

Gaylor, D.W. & Chen, J.J. (1986). Relative potency of chemical carcinogens in rodents. *Risk Analysis* 6, 283-290.

Gold, L.S., Sawyer, C.B., Mcgaw, R., Backman, G.M., de Veciana, M., Levinson, R., Hooper, N.K., Havender, W.R., Bernstein, L., Peto, R., Pike, M.C. & Ames, B.N. (1984). A carcinogenic potency database of the standardized results of animal bioassays. *Environmental Health Perspectives* 58, 9-319.

Gold, L.S., de Veciana, M., Backman, G.M., Mcgaw, R., Lopipero, P., Smith, M., Blumenthal, M., Levinson, R., Gevson, J., Bernstein, L. & Ames, B.N. (1986a). Chronological supplement to the carcinogenic potency database: standardized results of animal bioassays published through December 1982. *Environmental Health Perspectives* 67, 161-200.

Gold, L.S., Bernstein, L., Kaldor, J., Backman, G. & Hoel, D. (1986b). An empirical comparison of methods used to estimate carcinogenic potency in long-term animal bioassays: lifetable vs. summary incidence data. *Fundamental and Applied Toxicology* 6, 263-269.

Gold, L.S., Slone, T.H., Backman, G.M., Mcgaw, R.,DaCosta, M., Lopipero, P., Blumenthal, M. & Ames, B.N. (1987). Second chronological supplement to the carcinogenic potency database: standardized results of animal bioassays published through December 1984 and by the National Toxicology Program through May 1986. *Environmental Health Perspectives* 74, 237-329.

Gold, L.S., Slone, T.H. & Bernstein, L. (1989). Summary of carcinogenic potency and positivity for 492 rodent carcinogens in the carcinogenic potency database. *Environmental Health Perspectives* 79,

259-272.

Gold, L.L. & Ames, B.N. (1990). The importance of ranking possible carcinogenic hazards using HERP. *Risk Analysis.* In press.

Gold, L.S., Slone, T.H., Backman, G.M., Eisenberg, S., Da Costa, M., Wong, M., Manley, N.B., Rohrbach, L. & Ames, B.N. (1990). Third chronological supplement to the Carcinogenic Potency Database: Standardized results of animal bioassays published through December 1986 and by the National Toxicology Program through June 1987. *Environmental Health Perspectives* 84, 215-286.

Gold, L.S., Slone, T.H., Stern, B.R., Manley, N.B. & Ames, B.N. (1992). Rodent carcinogens: setting priorities. *Science* 258, 261-265.

Goodman, G., Shlyakhter, A. & Wilson, R. (1991). The relationship between carcinogenic potency and maximum tolerated dose is similar for mutagens and nonmutagens. In: Chemically Induced Cell Proliferation: *Implications for Risk Assessment* (Butterworth, B.B., Slaga, T.J., Garland, W. & McLean, M., eds.), Wiley-Liss, New York, pp. 501-516.

Goodman, G. & Wilson, R. (1991a). Predicting the carcinogenicity of chemicals in humans from rodent bioassay data. *Environmental Health Perspectives* 94, 195-218.

Goodman, G. & Wilson, R. (1991b). Quantitative prediction of human cancer risk from rodent carcinogenic potencies: a closer look at the epidemiological evidence for some chemicals not definitively carcinogenic in humans. *Regulatory Toxicology and Pharmacology* 14, 118-146.

Goodman, G. & Wilson, R. (1992). Comparison of the dependence of the TD_{50} on maximum tolerated dose for mutagens and nonmutagens. *Risk Analysis.* To appear.

Haseman, J.K. (1985). Issues in carcinogenicity testing: dose selection. *Fundamental and Applied Toxicology* 5, 66-78.

Haseman, J.K. & Huff, J.E. (1987). Species correlation in long-term carcinogenicity studies. *Cancer Letters* 37, 125-132.

Hoel, D.G., Haseman, J.K., Hogan, M.D., Huff, J. & McConnell, E.E. (1988). The impact of toxicity on carcinogenicity studies: implication for risk assessment. *Carcinogenesis* 9, 2045-2053.

International Agency for Research on Cancer (1987). Overall Evaluations of Carcinogenicity: An Updating of IARC Monographs Volumes 1 to 42. *IARC Monographs on the Evaluation of Carcinogenic Risks to Humans, IARC, Lyon.*

Kaldor, J.M., Day, N.E. & Hemminki, K. (1988). Quantifying the carcinogenicity of antineoplastic drugs. *European Journal of Cancer* 24, 703-711.

Kodell, R.L., Gaylor, D.W. & Chen, J.J. (1991). Carcinogenic potency correlations: real or artifactual? *Journal of Toxicology and Industrial Health*. 32, 1-9.

Krewski, D. (1990). Measuring carcinogenic potency. *Risk Analysis* 10, 615-617.

Krewski, D. & Van Ryzin, J. (1981). Dose response models for quantal response toxicity data. In: Statistics and Related Topics. *(M. Csorgo, D. Dawson, J.N.K. Rao & E. Saleh, eds.). North Holland, Amsterdam,* pp. 201-231.

Krewski, D., Murdoch, D. & Withey, J. (1989). Recent developments in carcinogenic risk assessment (with discussion). *Health Physics* 57 (Supplement 1), 313-326.

Krewski, D., Goddard, M.J. & Withey, J. (1990a). Carcinogenic potency and interspecies extrapolation. In: Mutation and the Environment, Part D: *Carcinogenesis (M.L. Mendelsohn & R.J. Albertini, eds.). Wiley-Liss, New York,* pp. 323-334.

Krewski, D., Szyszkowicz, M. & Rosenkranz, H. (1990b). Quantitative factors in chemical carcinogenesis: variation in carcinogenic potency. *Regulatory Toxicology and Pharmacology* 12, 13-29.

Krewski, D., Gaylor, D. & Szyszkowicz, M. (1991a). A model-free approach to low dose extrapolation. *Environmental Health Perspectives* 90, 279-285.

Krewski, D., Withey, J., Ku, L.F. & Travis, C.C. (1991b). Physiologically based pharmacokinetic models: applications in carcinogenic risk assessment. In: New Trends in Pharmacokinetics *(A. Rescigno & A. Thakur, eds.). Plenum, New York,* pp. 355-390.

Krewski, D., Leroux, B.G., Creason, J. & Claxton (1992). Sources of variation in the mutagenic potency of complex chemical mixtures based on the Salmonella/microsome assay. *Mutation Research* 276, 33-59.

Krewski, C., Leroux, B.G., Bleuer, S. & Broekhoven (1993). Modeling the Ames Salmonella/microsome assay. *Biometrics*. In press.

Lave, L.B., Ennever, F.K., Rosenkranz, H.S. & Omenn, G.S. (1988).

Information value of the rodent bioassay. *Nature(London)* 336, 631-633.

Louis, T.A. (1984). Estimating a population of parameter values using Bayes and empirical Bayes methods. *Journal of the American Statistical Association* 79, 393-398.

Lutz, W.K. (1990). Dose response relationship and low dose extrapolation in chemical carcinogenesis. *Carcinogenesis* 11, 1243-1247.

Mantel, N. & Bryan, W.R. (1961). Safety "testing of carcinogenic agents. *Journal of the National Cancer Institute* 27, 455-470.

McGregor, D.B. (1992). Chemicals classified by the IARC: their potency in rodent carcinogenicity tests, genotoxicity and acute toxicity. In: *Mechanisms of Carcinogenesis in Risk Identification (H. Vainio, P.N. Magee, D.B. McGregor & A.J. McMichael, eds.)*. International Agency for Research, Lyon. In press.

McCann, J., Gold, L.S., Horn, L., McGill, R., Graedel, T.E. & Kaldor, J. (1988). Statistical analysis of Salmonella test data and comparison of results of animal cancer tests. *Mutation Research* 205, 183-195.

McConnell, E.E. (1989). The maximum tolerated dose: the debate. *Journal of the American College of Toxicology* 8, 1115-1120.

Meselson, M. & Russell, K. (1977). Comparisons of carcinogenic and mutagenic potency. In: *Origins of Human Cancer, Book C: Human Risk Assessment (H.H. Hiatt, J.D. Watson & J.A. Winston, eds.)*. *Cold Spring Harbor Laboratory, Cold Spring Harbor*, pp. 1473-1481.

Metzger, B., Crouch, E. & Wilson, R. (1987). On the relationship between carcinogenicity and acute toxicity. *Risk Analysis* 9, 169-177.

Moolgavkar, S.H. & Luebeck, G. (1990). Two-event model for carcinogenesis: biological, mathematical and statistical considerations. *Risk Analysis* 10, 323-341.

Munro, I.C. (1977). Considerations in chronic toxicity testing: the chemical, the dose, the design. *Journal of Environmental Pathology and Toxicology* 1, 183-197.

Munro, I.C. (1990). Safety assessment procedures for indirect food additives: an overview. *Regulatory Toxicology and Pharmacology* 12, 2-12.

Murdoch, D.J. & Krewski, D. (1988). Carcinogenic risk assessment with time-dependent exposure patterns. *Risk Analysis* 8, 521-530.

Murdoch, D.J., Krewski, D. & Wargo, J. (1992). Cancer risk assessment with intermittent exposure. *Risk Analysis.* In press.

National Research Council (1992). Issues in Risk Assessment, Vol. 1. Use of the Maximum Tolerated Dose in Animal Bioassays for Carcinogenicity. *National Academy Press, Washington, D.C.*

Nesnow, S. (1990). A multifactor ranking scheme for comparing the carcinogenic activity of chemicals. *Mutation Research* 239, 83-116.

Oser, B. (1981). The rat as a model for human toxicological evaluation. *Journal of Toxicology and Environmental Health* 8, 521-542.

Parodi, S., Taningher, M. & Santi, L. (1982a). Alkaline elution in vivo: fluorometric analysis in rats. Quantitative predicition of carcinogenicity, as compared with other short-term tests. In: *Indicators of Genotoxic Exposure (B.A. Bridges, B.E. Butterworth & I.B. Weinstein, eds.). Banbury Report No. 13, Cold Spring Harbor Laboratory, Cold Spring Harbor*, pp. 137-155.

Parodi, S., Taningher, M., Boero, P. & Santi, L. (1982b). Quantitative correlations among alkaline DNA fragmentation, DNA covalent binding, mutagenicity in the Ames test, and carcinogenicity for 21 compounds. *Mutation Research* 93, 1-24.

Parodi, S., Taningher, M., Romano, P., Grilli, S. & Santi, L. (1990). Mutagenic and carcinogenic potency indices and their correlation. *Teratogenesis, Carcinogenesis, and Mutagenesis* 10, 177-198.

Peto, R., Pike, M.C., Bernstein, L., Gold, L.S. & Ames, B.N. (1984). The TD50: a proposed general convention for the numerical description of the carcinogenic potency of chemicals in chronic-exposure animal experiments. *Environmental Health Perspectives* 58, 1-8.

Piegorsch, W.W. & Hoel, D.G. (1988). Exploring relationships between mutagenic and carcinogenic potencies. *Mutation Research* 196, 161-175.

Piegorsch, W.W., Carr, G.J., Portier, C.J. & Hoel, D.G. (1992). Concordance of carcinogenic response between rodent species: potency dependence and potential underestimation. *Risk Analysis* 12, 115-121.

Portier, C.J. & Hoel, D.G. (1987). Issues concerning the estimation of the TD_{50}. *Risk Analysis* 7, 437-447.

Rao, C.R. (1973). Linear Statistical Inference and its Applications (second edition). *Wiley, New York.*

Reith, J.P. & Starr, T.B. (1989a). Experimental design constraints on carcinogenic potency estimates. *Journal of Toxicology and Environ-*

mental Health 27, 287-296.

Reith, J.P. & Starr, T.B. (1989b). Chronic bioassays: relevance to quantitative risk assessment of carcinogens. *Regulatory Toxicology and Pharmacology* 10, 160-173.

Rosenkranz, H.S. & Ennever, F.K. (1990). An association between mutagenicity and carcinogenic potency. *Mutation Research* 244, 61-65.

Rulis, A.M. (1986). De minimis and the threshold of regulation. In: *Food Protection Technology (C.W. Felix, ed.). Lewis Publishers, Chelsea, Michigan*, pp. 29-37.

Sawyer, C., Peto, R., Bernstein, L. & Pike, M.C. (1984). Calculation of carcinogenic potency from long-term animal carcinogenesis experiments. *Biometrics* 40, 27-40.

Shelby, M.D. (1988). The genetic toxicity of human carcinogens and its implications. *Mutation Research* 204, 3-15.

Schmidt-Nielson, K. (1984). Scaling. Why is Animal Size so Important? *Cambridge University Press, Cambridge.*

Shlyakhter, A., Goodman, G. & Wilson, R. (1992). Monte Carlo simulation of animal bioassays. *Risk Analysis* 12, 73-82.

Sweet, D.V. (ed.) (1987). Registry of Toxic Effects of Chemical Substances (RTECS) 1986 Edition. *National Center for Occupational Safety and Health, Washington, D.C.*

Tomatis, L., Aitio, A., Wilbourn, J. & Shuker, L. (1989). Human carcinogens so far identified. *Japanese Journal of Cancer Research* 80, 7995-807.

Travis, C.C. & Bowers, J.C. (1991). Interspecies scaling of anesthetic potency. *Toxicology and Industrial Health* 7, 2499-260.

Travis, C.C. & White, R.K. (1988). Interspecies scaling of toxicity data. *Risk Analysis* 8, 119-125.

Travis, C.C., Richter Pack, S.A., Saulsbury, A.W. & Yambert, M.W. (1990a). Prediction of carcinogenic potency from toxicology data. *Mutation Research* 241, 21-36.

Travis, C.C., Saulsbury, A.W. & Richter Pack, S.A. (1990b). Prediction of cancer potency using a battery of mutation and toxicity data. *Mutagenesis* 5, 213-219.

Travis, C.C., Wang, L.A. & Waehner, M.J. (1991). Quantitative correlation of carcinogenic potency with four different classes of short-term test data. *Mutagenesis* 6, 353-360.

carcinogen hazards using the TD_{50} (with discussion). *Risk Analysis* 10, 609-614.

U.S. Environmental Protection Agency (1986). Guidelines for carcinogen risk assessment. *Federal Register* 51, 33992-34003.

Van Ryzin, J. (1980). Quantitative risk assessment. *Journal of Occupational Medicine* 22, 321-326.

Williams, P. & Portier, C.J. (1992). The importance of mutagenicity and chemical structure in the shape of carcinogenicity dose response curves. *Submitted.*

Woodward, K.N., McDonald, A. & Joshi, S. (1991). Ranking of chemicals for carcinogenic potency - a comparative study of 13 carcinogenic chemicals and an examination of some of the issues involved. *Carcinogenesis* 12, 1061-1066.

Zeiger, E., Anderson, B., Haworth, S., Lawlor, T. & Mortelmans, K. (1988). Salmonella mutagenicity tests: IV. Results from the testing of 300 chemicals. *Environmental and Molecular Mutagenesis 11(Suppl. 12),* 1-158.

Zeise, L., Wilson, R. & Crouch, E.A.C. (1984). Use of acute toxicity to estimate carcinogenic risk. *Risk Analysis 4,* 187-199.

Zeise, L., Crouch, E.A.C. & Wilson, R. (1986). A possible relationship between toxicity and carcinogenicity. *Journal of the American College of Toxicology 5,* 137-151.

1. Health Protection Branch, Health & Welfare Canada, Ottawa, Ontario, Canada

2. Department of Mathematics & Statistics, Carleton University, Ottawa, Canada

3. National Center for Toxicological Research, Food & Drug Administration, Jefferson, Arkansas

4. Center for Mathematical Sciences, University of Wisconsin-Madison, Madison, Wisconsin

5. Department of Mathematical Sciences, University of Wisconsin-Milwaukee, Milwaukee, Wisconsin

**Annex A: Maximum Likelihood Methods
for Fitting the Weibull Model**

Suppose that the probability $P(d)$ of a tumor occurring at dose d follows the Weibull model

$$P(d) = 1 - \exp\{-(a+bd^k)\} \qquad (A.1)$$

$(a,b,k > 0)$ as in (2.5). We wish to estimate the unknown model parameters a, b and k on the basis of an experiment with $s+1$ dose levels $0 = d_o < d_1 < .... < d_s$. Suppose that x_i of the n_i animals in group $i = 0,1,...,s$ develop tumors. Estimators of the unknown model parameters may be obtained by maximizing the binomial likelihood

$$L(a,b,k|x) = \prod_{i=0}^{s} \binom{n_i}{x_i} p_i^{x_i}(1-p_i)^{n_i-x_i} \qquad (A.2)$$

where $p_i = P(d_i)$ and $x = (x_0, x_1 ..., x_k)$. Numerical procedures for obtaining the maximum likelihood estimators (mle's) of the unknown model parameters, as well as the mle of the TD_{50} and its standard error, are described by Krewski & Van Ryzin (1981).

It is possible that this likelihood may not attain a global maximum, in which case the mle's of the unknown parameters do not exist. To illustrate, take $s = 2$, $n_0 = n_1 = n_2 = n$, and suppose that $x_0 = x_1 \equiv x$ with $x_2 \equiv y > x$. The likelihood function L then satisfies the upper bound

$$[\binom{n}{x}\left(\frac{x}{n}\right)^x\left(1-\frac{x}{n}\right)^{n-x}]^2\binom{n}{y}\left(\frac{y}{n}\right)^y\left(1-\frac{y}{n}\right)^{n-y}]=L^*. \qquad (A.3)$$

Let c_0 and c_1 be defined by the equations

$$\frac{x}{n} = 1-\exp[-c_0] \qquad (A.4)$$

and

$$\frac{y}{n} = 1 - \exp[-(c_0 + c_1)].$$ (A.5)

If $k \to \infty$ and $b \to 0$ or ∞ with $bd_2^k = c_1$ held constant, then $bd_1^k = c_1(d_1/d_2)^k \to 0$ and $L \to L^*$. Thus, no finite mle of k exists in this case. This seems intuitively reasonable, since data of the type under consideration are consistent with dose response curves of arbitrarily large upward curvature (i.e., arbitrarily large values of k). Noting that

$$TD_{100p} = \left[-\frac{\log_e(1-p)}{b} \right]^{1/k},$$ (A.6)

$(0 < p < 1)$, it follows that the mle of the TD_{100p} is equal to d_2 for any value of p in this case, an unpleasant conclusion. Other estimation methods such as least squares may be expected to perform in a similar manner.

Of the 217 data sets considered by Krewski et al. (1990b), mle's were readily obtained for the 122 dose response curves that were strictly increasing. The mle's for a further 69 data sets did not appear to exist because of nonmonotonicity as discussed above. The final 26 data sets involved only a control group and single nonzero dose, so that the shape parameter k could not be estimated.

For the 122 data sets for which mle's could be obtained, an adjusted measure of carcinogenic potency given by

$$TD_{50}^* = f^{2/k} TD_{50}$$ (A.7)

was calculated using the factor $f^{2/k}$ discussed in annex C. This effectively adjusts all TD_{50} values to a two-year standard rodent lifespan. By linear approximation (Rao, 1973), the variance of

$$\log_{10}TD*_{50} = \frac{1}{k}\left[\log_{10}[f^2(\log_e 2)] - (\log_{10}b)\right]$$ (A.8)

is given by

$$V(\log_{10}TD_{50}^{*}) = k^{-4}\log_{10}^{2}(\frac{f^{2}\log_{e}2}{b})V(k) + k^{-2}V(\log_{10}b)$$

$$+ 2k^{-3}\log_{10}e\log_{10}(\frac{f^{2}\log_{e}2}{b})Cov(\log_{10}b,k).$$

(A.9)

Estimates of $V(k)$, $V(\log_{10}b)$ and $Cov(\log_{10}b, k)$ can be obtained using RISK 81 (Krewski and Vany Ryan, 1981).

Rather than discard the 69 data sets for which mle's could not be obtained, we chose to fit a Weibull model to each of these data sets using a fixed value of the shape parameter k. In this regard, we first separated the 69 data sets into two subgroups based on their overall shape. A value of $k = 1.7$ was used for the 42 data sets that demonstrated clear upward curvature, this being the median value of k observed among the 68 of the 122 data sets for which $k > 1$. Similarly, a value of $k = 0.55$ was used for the 27 data sets exhibiting downward curvature, this being the median value of the 54 of the 122 data sets for which $k < 1$. The variance of $\log_{10}TD_{50}^{*}$ was then estimated using (A.9), with k treated as an estimated rather than a known parameter. Allowance for some degree of uncertainty in the value of k is desirable in order not to severely underestimate the variance of $\log_{10}TD_{50}^{*}$ (cf. annex B).

The 26 data sets in which only a control and single dose group were available were not used here since no information on the shape of the dose response curve is available.

Annex B. Shrinkage Estimators of the Distribution of Carcinogenic Potency

The distribution of TD_{50} values for a series of chemical carcinogens provides useful information on the variation in carcinogenic potency. Because each estimate $\hat{TD}_{50}$ of the true TD_{50} for a specific chemical is subject to estimation error, the distribution of estimated potency values ($\hat{TD}_{50}$s) will exhibit greater dispersion than the distribution of *true* potency values (TD_{50}s). This overdispersion may be eliminated using empirical Bayes shrinkage estimators (Louis, 1984).

Let $Y = \log_{10}TD_{50}$ and suppose that $E(Y) = \mu = \log_{10}TD_{50}$, with $V(Y) = \sigma^2$. Let $Y_1,...,Y_n$ denote the logarithms of the estimated TD_{50} values for a series of n chemical carcinogens. We suppose that Y_i is normally distributed with mean μ_i and variance σ_i^2. We further suppose that μ_i are normally distributed with mean μ and variance τ^2 where τ^2 reflects the variance among the μ_i. Our objective is to estimate μ and τ^2, and hence describe the lognormal distribution of unknown TD_{50} values.

Noting that

$$E(Y_i - \mu)^2 = E(\sigma_i^2) + \tau^2, \tag{B.1}$$

an estimator of τ^2 is

$$\hat{\tau}^2 = \sum_{i=1}^{n} \frac{(Y_i - \overline{Y})^2}{n-1} - \sum_{i=1}^{n} \frac{\hat{\sigma}_i^2}{n}, \tag{B.2}$$

where $\overline{Y} = \sum Y_i/n$ and $\hat{\sigma}_i^2$ is the estimator of $V(\log_{10}TD_{50})$ based on (A.9).

The shrinkage estimator of μ_i is given by

$$\hat{\mu}_i = \hat{\mu} + \hat{F}\hat{D}_i(Y_i - \hat{\mu}), \tag{B.3}$$

where $\hat{D}_i = \hat{\tau}^2/(\hat{\tau}^2 + \sigma_i^2)$ represents an estimator of the intrastudy correlation, $\hat{\mu} = \sum \hat{D}_i Y_i / \sum \hat{D}_i$ is an estimator of the overall mean of the log potency distribution, and

$$\hat{F}^2 = 1 + \frac{n-1}{n} \frac{\sum \hat{D}_i \hat{\sigma}_i^2}{\sum [\hat{D}_i(Y_i - \hat{\mu})]^2} \tag{B.4}$$

is designed to protect against overadjustment for overdispersion. In general $\hat{F}\hat{D}_i < 1$, so that the estimators $\hat{\mu}_i$ of the μ_i are obtained by "shrinking" the Y_i toward the mean $\hat{\mu}$. The estimators $\hat{\mu}_i$ of μ_i have the correct dispersion in that

$$E\left[\sum \frac{(\hat{\mu}_i - \hat{\mu})^2}{n-1}\right] \approx \tau^2. \tag{B.5}$$

In fitting the Weibull model in (A.1), we found that the estimate of the variance of the $\log_{10}TD_{50}^*$ based on (A.9) appeared to be excessively large in a small number of cases. In order not to underestimate the between-study variability $\hat{\tau}^2$ based on (B.2), we used a trimmed mean $\Sigma^* \sigma_i^2/n^*$, in which the largest and smallest 10% of the observed values of σ_i^2 were omitted (Hampel et al., 1986, p. 79). Specifically, the summation Σ^* covers only those $n^* = 153$ observations falling in the central 80% of the distribution of the σ_i^2.

Annex C: Adjustment of Potency Values for Less than Lifetime Exposure

In order to ensure that TD_{50} values for different chemicals are comparable, some adjustment for differences in the duration of the experimental period is desirable. Gold et al. (1984) adjusted TD_{50} values by a multiplicative factor of f^2, where f represents the fraction of a two-year period encompassed by the study period. This effectively scales the TD_{50} values to a standard two-year rodent lifetime. Specifically, we have

$$TD_{50}^* = f^2 TD_{50}, \tag{C.1}$$

where the TD_{50} denotes the estimate of carcinogenic potency based on the observed data for the actual experimental period, and denotes the standardized value.

To motivate the use of the adjustment factor f^2, consider the extended Weibull model

$$P(d,t) = 1 - \exp\{-(a+bd^k)t^p\} \tag{C.2}$$

depending on both dose d and time t. Under this model, the $TD_{50}(t)$ evaluated at time t is given by

$$TD_{50}(t) = \left(\frac{\log_e 2}{bt^p}\right)^{1/k} .$$ (C.3)

Thus, the ratio of TD_{50}'s at two distinct times t_1 and t_2 is

$$\frac{TD_{50}(t_2)}{TD_{50}(t_1)} = \left[\left(\frac{t_1}{t_2}\right)^p\right]^{1/k} = f^{p/k},$$ (C.4)

where $f = t_1/t_2$. In the CPDB, Gold et al. (1984) use a one-stage model with k=1 and set p=2 based on empirical observations reported by Peto et al. (1984), leading to their adjustment factor f^2. In our applications of the Weibull model in (A.1), we will use a similar adjustment factor of $f^{2/k}$ to standardize TD_{50} values to a two-year rodent lifetime.

For a multi-stage model of the form

$$P(d,t) = 1 - \exp\{-(q_0 + q_1 d + \cdots + q_k d^k)t^p\}$$ (C.5)

allowing for the effects of both dose d and time t, the TD_{50} at time t is obtained as the solution of the equation

$$q_1[TD_{50}(t)] + \cdots + q_k[TD_{50}(t)]^k = \frac{\log_e 2}{t^p}.$$ (C.6)

It follows that the standardized value of the TD_{50} is obtained as the solution of the equation

$$q_1[TD_{50}^*] + \cdots + q_k[TD_{50}^*]^k = \frac{\log_e 2^{f^p}}{t^p}.$$ (C.7)

As with the Weibull model, we set p=2 in the applications considered in this paper.

Annex D. Correlation Between
TD_{50} and MTD

In this annex, we derive an analytical expression for the correlation between the TD_{50} and the MTD. To this end, suppose that the probability $P(d)$ of a tumor occurring in an animal exposed to dose $D = $ MTD satisfies the Weibull model

$$P(D) = 1 - \exp \{-(\alpha + \beta D^k)\} \qquad (D.1)$$

in (2.5), where the background parameter $\alpha > 0$ and the shape parameter $k > 0$ are known. This is a generalization of the one-stage model used by Bernstein et al. (1985) in which $k = 1$.

Suppose that x of the n animals exposed to dose D develop tumors. Since α and k are assumed known, β may be estimated by

$$\hat{\beta} = D^{-k} [-\log_e \left(\frac{1 - \frac{x}{n}}{1 - p_0} \right)], \qquad (D.2)$$

where $p_0 = P(0)$ describes the spontaneous response rate. This leads to an estimate

$$\hat{TD}_{50} = \left(\frac{\log_e 2}{\hat{\beta}} \right)^{1/k}. \qquad (D.3)$$

of the TD_{50}.

The estimate of β is appropriate for $r \leq x \leq n-1$. The lower limit of $x = r$ is the minimum value of x that would lead to a statistically significant result at a nominal significance level of $0 < \gamma < 1$; the value of r is determined from the fact that in the absence of a treatment effect at dose D, x follows the binomial distribution Bin $(50, P(D))$. The upper limit of $x = n-1$ is included since β, and hence TD_{50}, is undefined for $x = n$.

The constraint $x \leq n-1$ implies that

$$\hat{T}D_{50} \geq D \left[\frac{\log_e 2}{-\log_e\left(\frac{1-\frac{n-1}{n}}{1-p_0}\right)} \right]^{1/k} = Dg_1 = a,$$

(D.4)

whereas $x \leq r$ implies

$$\hat{T}D_{50} \leq D \left[\frac{\log_e 2}{-\log_e\left(\frac{1-\frac{r}{n}}{1-p_0}\right)} \right]^{1/k} = Dg_2 = b.$$

(D.5)

We wish to find the correlation between $Y = \log_e TD_{50}$ and $X = \log_e D$. (Although the correlation will be identical using logarithms to the base 10, the derivation of the correlation given here is simpler using natural logarithms.) Suppose now that $W = TD_{50}$ follows a uniform distribution on the interval [a,b], reflecting the fact that given the value D of the MTD, the estimated value of the TD_{50} is unrelated to the MTD. Suppose further that X follows some distribution with mean μ and variance σ^2. Although Bernstein et al. (1985) observed that the empirical distribution of X is approximately normal, the correlation between Y and X does not depend on the distribution of X other than through its variance σ^2.

To calculate corr (Y,X), note that

$$E(Y|X) = \frac{1}{b-a} \int_a^b \log_e w \, dw = h_1 - 1 + \log_e D$$

(D.6)

and

$$E(Y^2|X) = \frac{1}{b-a} \int_a^b (\log_e w)^2 \, dw$$
$$= h_2 - 2h_1 + 2 + 2(h_1 - 1) \log_e D + (\log_e D)^2,$$

(D.7)

where

$$h_1 = \frac{g_2 \log_e g_2 - g_1 \log_e g_1}{g_2 - g_1} \qquad \text{(D.8)}$$

and

$$h_2 = \frac{g_2 (\log_e g_2)^2 - g_1 (\log_e g_1)^2}{g_2 - g_1}. \qquad \text{(D.9)}$$

Thus we have

$$V(Y) = E[V(Y|X)] + V[E(Y|X)] = h_2 - h_1^2 + 1 + \sigma^2, \qquad \text{(D.10)}$$

with $V(X) = \sigma^2$. Noting that

$$E(XY) = E[XE(Y|X) = \mu(h_1 - 1) + \sigma^2 + \mu^2, \qquad \text{(D.11)}$$

where $\mu = E(X)$, we have

$$Cov(X,Y) = \sigma^2. \qquad \text{(D.12)}$$

This leads to the desired result:

$$\rho = \frac{Cov(X,Y)}{[V(X)V(Y)]^{1/2}} = [1 + \frac{h_2 - h_1^2 + 1}{\sigma^2}]^{-1/2}. \qquad \text{(D.13)}$$

It can be shown that $h_2 - h_1^2 + 1 \geq 0$, so that $0 < \rho \leq 1$. It can also be shown that $\rho \downarrow [\sigma^2/(\sigma^2 + 1)]^{1/2}$ as $k \downarrow 0$, and that $\rho \uparrow 1$ as $k \to \infty$. Thus $[\sigma^2/(\sigma^2 + 1)]^{1/2} \leq \rho \leq 1$. In the limiting case as $n \to \infty$, (D.13) reduces to

$$\rho = \left(\frac{\sigma^2}{\sigma^2 + 1} \right)^{1/2}. \qquad \text{(D.14)}$$

The values of the correlation coefficient ρ in (D.13) as a function of the sample size n are shown in Table 1. (Note that the values of h_1 and h_2 are implicit functions of n.) These results are based on a one-stage model (k = 1) with a spontaneous response rate p_0 = 0.10, and a nominal significance level of γ = 0.05 with r = 10 in the case n = 50. The value of σ^2 = V($\log_e$MTD) = 8.196 is based on the variance of the MTD of the 191 experiments considered previously by Krewski et al. (1990b). Using common logarithms, V($\log_{10}$MTD) = 1.546.

The dependency of the correlations between $\log_{10}TD_{50}$ and $\log_e$MTD on the Weilbull shape parameter k is illustrated in Table 2 for a sample size of n=50. These results, including the limiting cases as k $\rightarrow$ 0 or ∞, are also based on (D.13). Note that the correlation remains high regardless of the value of k.

Annex E: Correlation Between TD_{50}s For Rats and Mice

In this annex, we derive analytical expressions for the correlation between TD_{50} values for rats and mice. Letting Y_{rats} and Y_{mice} denote the logarithms (base^e) of the estimated TD_{50}s for rats and mice, we seek an expression for ρ = Corr(Y_{rats}, Y_{mice}). Following Bernstein et al. (1985) we assume initially that the MTD for rats is directly proportional to that for mice, with

$$MTD_{rats} = c\ MTD_{mice}. \qquad (E.1)$$

Using the notation of annex D, we will denote the logarithms of the MTDs for rats and mice by X_{rats} and X_{mice}, so that

$$X_{rats} = \log_e(c) + X_{mice}. \qquad (E.2)$$

Note that (E.2) implies that V(X_{rats}) = V(X_{mice}) = σ^2

As in annex D, we assume that the TD_{50}s for rats and mice are uniformly distributed about their respective MTDs. From (D.10), we may then write

$$V(Y_{rats}) = V(Y_{mice}) = h_2 - h_1^2 + 1 + \sigma^2, \qquad (E.3)$$

where h_1 and h_2 are the same for rats and mice since g_1 and g_2 defined in (D.4) and (D.5) respectively are the same for rats and mice. Assuming that Y_{rats} and Y_{mice} are conditionally independent, given MTD_{mice} (and hence MTD_{rats} from (D.1)), we have

$$\text{Cov } (Y_{rats}, Y_{mice}) =$$
$$\text{Cov } (E[Y_{rats}|TD_{rats}], E[Y_{mice}|MTD_{mice}]) = \sigma^2. \qquad \text{(E.4)}$$

Hence

$$\text{Corr}(Y_{rats}, Y_{mice}) = \left[1 + \frac{h_2 - h_1^2 + 1}{\sigma^2}\right]^{-1} = \rho^2, \qquad \text{(E.5)}$$

where $\rho = \text{Corr}(Y_{rats}, X_{rats})$ is given in (D.13) of annex D. Based on the $n = 127$ compounds from the CPDB considered in section 6.1, we find $\sigma^2_{rats} = 10.065 \approx \sigma^2_{mice} = 8.873$. For $\sigma^2 = 10$, we have $\rho = 0.943$.

The assumption (D.1) of strict proportionality between MTD_{rats} and MTD_{mice} can be relaxed. Let $V(X_{rats}) = \sigma^2_{rats}$ and $V(X_{mice}) = \sigma^2_{mice}$. As in (D.3), we have

$$V(Y_{rats}) = h_2 - h_1^2 + 1 + \sigma^2_{rats} \qquad \text{(E.6)}$$

and

$$V(Y_{mice}) = h_2 - h_1^2 + 1 + \sigma^2_{mice}. \qquad \text{(E.7)}$$

Assuming Y_{rats} and Y_{mice} are conditionally independent, given X_{rats} and X_{mice}, we have

$$\text{Cov}(Y_{rats}, Y_{mice}) = \text{Cov}(X_{rats}, X_{mice}), \qquad \text{(E.8)}$$

and hence

$$\text{Corr}(Y_{\text{rats}}, Y_{\text{mice}}) = \frac{\text{Cov}(X_{\text{rats}}, X_{\text{mice}})}{[(h_2 - h_1 + 1 + \sigma_{\text{rats}}^2)(h_2 - h_1^2 + 1 + \sigma_{\text{mice}}^2)]^{1/2}}$$

$$= \text{Corr}(X_{\text{rats}}, X_{\text{mice}})\text{Corr}(Y_{\text{rats}}, X_{\text{rats}})\text{Corr}(Y_{\text{mice}}, X_{\text{mice}}).$$

(E.9)

For the n = 127 compounds considered in section 6.1, we estimate Cov $(X_{\text{rats}}, X_{\text{mice}})$ = 7.638, and ρ = 0.763

Appendix G

Informal Search for "Supercarcinogens"

As discussed in the text of the report, the graph of TD_{50} versus MDT can be divided empirically into three regions (Figure 1). By observation, most carcinogens are in a narrow band (Region B); few are above (Region A) or below (Region C) this band. The absence of carcinogens from Region A might be partly or entirely artifactual. Any carcinogen whose biologic properties would place it in Region A would have such low potency (such a high TD_{50}) that it would give a very weak response even if tested at the MTD; hence, it would not be recognized as carcinogenic and would not be included in the CPDB. Thus, no conclusions can be drawn from the apparent emptiness of Region A.

The absence of carcinogens from Region C, however, is not obviously artifactual. The argument put forward by Rieth and Starr (1989b) that any carcinogens that belong properly in Region C would yield 100% tumor incidence in a conventional bioassay, so a finite TD_{50} could not be calculated, is incorrect. For some chemicals that yield 100% tumor incidence, the CPDB includes a 99% upper confidence limit on the TD_{50} (Gold et al., 1986a); for others, tumor incidence less than 100% can be observed in bioassays conducted at lower doses or for periods shorter than a lifetime. Thus, if some chemicals truly belong in Region C, they should be detectable, and it should be possible to derive numerical estimates of potency for at least some of them.

Nevertheless, the committee identified several types of bias that might

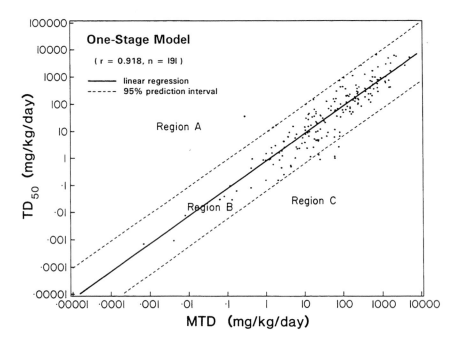

FIGURE 1

hypothetically have led to the exclusion of carcinogens that properly belong in Region C from the CPDB and from studies based on the CPDB. For purposes of discussion, we refer to these hypothetically excluded carcinogens that properly belong in Region C as "supercarcinogens"—defined for this purpose as carcinogens whose TD_{50}s lie below the 95% error bound on the regression line in Figure 1, i.e., less than about MTD/7. One hypothetical source of bias is that some of these agents were identified as potent carcinogens long ago and, being well known as such, were never tested in up-to-date bioassays and so did not have met the inclusion criteria of the CPDB. Another possible source of bias is that some of the agents, if tested at the MTD, yielded tumors in very short periods and were excluded from the CPDB because of early termination of the studies. Krewski's criteria for selecting chemicals from the CPDB for analysis could have introduced other, more subtle biases.

To investigate whether those hypothetical biases are important, the committee conducted a search for supercarcinogens that exist but have been excluded from the CPDB or from Krewski's analysis. The search was necessarily informal, because there is no systematic compilation of carcinogenic potencies other than the CPDB. The committee's approach was to compile a list of candidate chemicals with various criteria and then to review the data on them to explore whether they might fall into Region C, either according to the inclusion criteria and calculation procedures of the CPDB and of Krewski or according to modified criteria and procedures. The results of the search are reported in this appendix.

CRITERIA AND CANDIDATE CHEMICALS

The following criteria were used to identify candidate chemicals for this study:

A. "Classical" carcinogens—identified before 1965 and not subjected to modern bioassays.

B. Agents that induced tumors in less than 6 months and might never have been tested in a lifetime bioassay.

C. Other agents that are generally recognized as "potent" carcinogens and might never have been formally tested for carcinogenicity.

D.   Agents that have been tested over an unusually wide range of doses and are believed to be effective at doses below the MTD by a factor of at least 100.

E.   Other agents nominated by committee members.

On the basis of those criteria, the committee selected 18 candidate agents for study (Table G-1).

TABLE G-1   Chemicals and Other Agents Considered in this Study

Agent	Criteria for Inclusion in the Study[a]
2-Acetylaminofluorene	D
Acrylonitrile	E
Benzidine	B (parent compound of benzidine dyes)
Benzo[a]pyrene	A
1,3-Butadiene	D
Carbon tetrachloride	C
C.I. Direct Black 38	B
C.I. Direct Blue 6	B
C.I. Direct Brown 95	B
Dibenz[a,h]anthracene	A
Dimethyl sulfate	C
Ethylene dibromide	B
Ethylene oxide	E
Ethylnitrosourea	C
Methyl bromide	B
MOCA	E
Plutonium	A,B,C,D, (most potent member of class of radionuclides)
Vinyl Chloride	D

[a]See text.

DATA

Twelve of the candidates were already included in the CPDB and in Krewski's study (upper portion of Table G-2). For two other agents (benzo[a]pyrene and 1, 3-butadiene), the committee identified dose--response data that could be analyzed quantitatively (Tables G-3 and G-4). For another agent (vinyl chloride), the committee identified an ingestion study that gave results (Table G-5) markedly different from those of the inhalation study included in the first part of Table G-2. The ingestion study (Feron et al., 1981) appears to have met the inclusion criteria of the CPDB, and it is not clear why it was not included in the CPDB. The data in Tables G-3, G-4, and G-5 were analyzed by Krewski with the same methods as those used in his workshop paper, and the resulting estimates of TD_{50} are tabulated in the lower portion of Table G-2.

For four agents listed in Table G-1, comparable numerical estimates of carcinogenic potency could not be obtained, for the following reasons:

• *Dimethyl sulfate.* The only reported studies are unsuitable for quantitative analysis, but show tumors at the MDT and MDT/2 (IARC, 1974).

• *Dibenz[a,h]anthracene.* The only reported studies are unsuitable for quantitative analysis (ATSDR, 1990).

• *Methyl bromide.* Data purporting to show induction of forestomach tumors within 90 days (Danse et al., 1984) have been discredited (EPA, 1986; Reuzel et al., 1991).

• *Plutonium.* Dose data on this and other radionuclides are not commensurable with those customarily applied to chemical carcinogens. For plutonium, the radiation dose that causes early death (within 1.5 years) due to radiation pneumonitis and pulmonary fibrosis in animals exposed by inhalation is about 45 Gy (Scott et al., 1990), whereas the TD_{50} for animals similarly exposed is 3.3 Gy (Diehl et al., 1992). (In this case, early death is used as the measure of toxicity for the purpose of determining the MTD.)

TABLE G-2 Deviations from Krewski's Regressions

Krewski's Case No.	Chemical	MTD	log MTD	Gold's Estimate (one-stage, time-corrected)					Krewski's Estimate (one-stage, time corrected)				
				TD50	log TD50	Predicted log TD50	Difference	Standardized Difference	TD50	log TD50	Predicted log TD50	Difference	Standardized Difference
2	2-Acetyl-amino-fluorene	10.0	1.00	3.78	0.577	0.835	-0.318	-0.582	3.83	0.583	0.895	-0.312	-0.572
3	Acrylonitrile	5.69	0.755	5.31	0.725	0.647	0.078	0.144	5.61	0.755	0.647	0.102	0.187
1	Benzidine	80.0	1.903	8.99	0.954	1.811	-0.857	-1.571	9.10	0.959	1.811	-0.851	-1.561
3	C.I. Direct Black 38	60.0	1.778	0.945	-0.025	1.684	-1.708	-3.132	0.99	-0.002	1.684	-1.688	-3.095
6	C.I. Direct Blue 6	60.0	1.778	1.18	0.072	1.684	-1.612	-2.955	1.17	0.068	1.684	-1.616	-3.962
8	C.I. Direct Brown 95	75.0	1.875	2.07	0.316	1.782	-1.466	-2.688	2.62	0.418	1.782	-1.364	-2.500
3	CCl_4	1650	3.217	114	2.057	3.143	-1.086	-1.991	115.12	2.061	3.143	-1.082	-1.983
3	EDB	28.1	1.449	1.26	0.100	1.350	-1.250	-2.291	5.60	0.748	1.350	-0.608	-1.103
1	ENU	0.429	-3.68	0.904	-0.044	-0.491	0.447	0.820	0.94	-0.026	-0.491	0.464	0.851
2	Ethylene	6.11	0.786	7.43	0.871	0.678	0.193	0.354	7.69	0.786	0.678	0.208	0.381
9	MOCA	34.0	1.531	20.8	1.318	1.434	-0.116	-0.212	21.07	1.324	1.434	-0.110	-0.202
13	Vinyl chloride (CPDB)	0.279	-0.554	14.2	1.152	-0.681	1.833	3.360	33.52	1.525	-0.681	2.206	4.044

TABLE G-2 Continued.

Krewski's Case No.	Chemical	MTD	log MTD	Gold's Estimate (one-stage, time-corrected)				Krewski's Estimate (one-stage, time corrected)				
				log TD50	Predicted log TD50	Difference	Standardized Difference	TD50	log TD50	Predicted log TD50	Difference	Standardized Difference
Chemicals or studies not in the CPDB or in Krewski's analysis												
	Benzo[a]-pyrene	32.5	1.512					11.70	1.068	1.414	-0.346	-0.634
	1,3-Butadiene (δ)	380	2.580					20.81	1.318	2.496	-1.178	-2.160
	Vinly Chloride (CRAM)	14.1	1.149					4.89	0.689	1.046	-0.357	-0.654

TABLE G-3   1,3 Butadiene*

	Dose Rate (mg/kg-d)	Tumor Incidence	
		Males	Females
Lymphocytic	0	2/70	2/70
lymphoma	3.8	1/70	4/70
	12	2/70	6/70
	38	4/70	3/70
	120	2/70	11/70
	380	62/90	36/90

*Inhalation exposure, 6h/day, 5d/wk for up to 2 years. Most animals died in high exposure groups by 65 weeks because of high tumor incidence.

Source: Melnick et al., 1990.

TABLE G-4   Benzo[a]pyrene*

	Dose Rate (mg/kg-d)	Tumor Incidence Male and Female
Stomach,	0	0/289
squamous cell carcinomas	0.13	0/25
and papillomas	1.3	0/24
	2.6	1/23
	3.9	0/37
	5.2	1/40
	5.85	4/40
	6.5	24/34
	13.0	19/23
	32.5	66/73

*Oral exposure in diet. Mice, CFW, male and female. Duration of exposure: 110 days. Duration of experiment: 183 days.

Source: Neal and Rigdon, 1967.

TABLE G-5   Vinyl Chloride*

	Dose Rate (mg/kg-d)	Tumor Incidence	
		Males	Females
Liver tumors	0	0/55	2/57
(neoplastic nodules,	1.7	2/58	26/58
hepatocellular carcino-	5.0	17/56	42/59
mas, angiosarcomas)	14.1	58/59	56/57

*Oral lifetime exposure. Surviving males and females were sacrificed at 135 and 144 weeks, respectively.

Source: Feron et al., 1981.

RESULTS

For the 14 candidate agents on which comparable quantitative data are available, the right side of Table G-2 shows the observed TD_{50} as calculated by Krewski's procedures with the one-stage model. The last three columns in Table G-2 show the $\log_{10} TD_{50}$ predicted by Krewski's model, the deviation from the regression line (observed - predicted $\log_{10} TD_{50}$), and the standardized deviation (observed deviation divided by r.m.s. error). The data are plotted in relation to Krewski's regression line in Figure 1. Most of the 14 candidate chemicals are within the 95% confidence limits (standardized deviation, 1.96); this is illustrated in Figure 2, which plots the calculated TD_{50}s for each of the 14 chemicals. For five of the 14 agents (the three benzidine dyes, carbon tetrachloride, and 1,3-butadiene), the calculated TD_{50}s are below the lower confidence limit on the regression line, i.e., inside Region C (Figure 2). Among the four agents on which comparable quantitative data are not available, only plutonium has a low ratio of TD_{50} to HDT (1:13), but its HDT caused premature deaths and would not be accepted as an MTD in a conventional bioassay.

Neither of the TD_{50}s calculated for vinyl chloride fall in Region C. The TD_{50} based on the data selected by the committee falls below the regression line in Region B; the TD_{50} based on the data in the CPDB falls above the upper confidence limit on the regression line in Region A.

The TD_{50}s in the CPDB for the 12 potential supercarcinogens already included in Krewski's 191 chemicals are generally very close to those calculated by Krewski. There are, however, two notable differences: EDB and vinyl chloride. The TD_{50} for EDB in the CPDB is based on life-table methods, and is thus somewhat different than that calculated by Krewski using summary tumor incidence data. The discrepency between the Gold and Krewski TD_{50}s for vinyl chloride based on the data from the CPDB is apparently due to differences in the numerical procedures used in model fitting. (This difference is small in relation to the wide variation in TD_{50}s in the CPDB based on different experiments with vinyl chloride.) Neither of these differences is particularly relevant to the search for supercarcinogens because the $TD50$s for these two compounds do not fall in Region C.

DISCUSSION

The results just discussed do not provide strong evidence of the existence of supercarcinogens. Of the 14 chemicals considered as potential supercarcinogens, only five fall inside Region C; even these five are only slightly beyond the boundary separating regions B and C.

These results are based on certain assumptions about the appropriate adjustments to be applied to dose (and hence to potency) in experiments that are terminated substantially earlier than the 2-year lifetime of rodents. Those assumptions are based on sparse empirical evidence and are somewhat arbitrary. A common generalization is that cancer incidence is proportional to t^n where the exponent n may range from 2 to 6 (Armitage and Doll, 1961). The CPDB's procedures are equivalent to the assumption that n = 2, which gives relatively low estimates of carcinogenic potency. An assumption that n = 3 or higher would shift the estimates of TD_{50} for the chemicals under review still further into Region C.

In summary, the results of the committee's informal study suggest that supercarcinogens are rare. The best candidates for designation as supercarcinogens are a few agents that induce cancer in rodents unusually early in life. For such agents, the definition of potency is somewhat arbitrary: the more account that is taken of their early action, the higher the estimates of potency and the weaker the general relationship between potency and toxicity.

Issues in Risk Assessment

The Two-Stage Model
Of Carcinogenesis

The Two-Stage Model
of Carcinogenesis

INTRODUCTION

In recent years, it has become clear that carcinogenesis is a multistep process that requires deregulation of cellular growth. Cell growth and differentiation are normally under genetic regulation, so it may be assumed that the critical events in carcinogenesis involve genetic damage and inappropriate genetic expression (Weinberg, 1988; 1989). Mathematical models based on those biologic considerations can be simple or complex depending on assumptions about the number and nature of the events required to transform normal cells into cancer cells and about the sequence of events.

The simplest model judged to be consistent with the data available—a model that assumes two critical stages—was selected for evaluation by the Committee on Risk Assessment Methodology (CRAM). The two-stage model has been proposed as an improvement over currently used models for estimating carcinogenic risks to health, because it incorporates biologic considerations, notably cell population kinetics. The principal purposes of this CRAM study were to assess the scientific basis of the two-stage model of carcinogenesis and to evaluate the possible applications of the two-stage model to health risk assessment.

As part of the information-gathering process, the committee held a workshop on November 8, 1990, with presentations by the originators and proponents of the two-stage model and by invited discussants. A workshop summary appears in Appendix A. Postworkshop discussions were also held with the workshop speakers and with representatives of the federal liaison group to clarify the issues. However, this report and its recommendations were prepared solely by the committee.

BIOLOGIC CONSIDERATIONS

Clinicians and pathologists have long recognized that cancer formation in humans is often preceded by a series of preneoplastic changes. Confirmation of the multistage nature of certain human cancers has been obtained by studies of the role of changes in oncogenes and suppressor genes in human colon cancer (Hollstein et al., 1991). Similar observations have been made on laboratory animals that were exposed to carcinogens experimentally (Barbacid, 1987; Balmain and Brown, 1988). Morphologic or histopathologic studies do not always lend themselves well, however, to conclusions as to the biologic potential or ultimate fate of individual precancerous lesions. Some uncertainty exists about the identification of particular lesions as part of a neoplastic process, their place in the pathologic sequence, the inevitability of their progression to the next stage, and the rate of transition when they progress. For risk assessment, it is important to be able to distinguish lesions that are reversible from lesions that will irreversibly lead to neoplastic disease. The frequency with which most preneoplastic lesions pass from stage to stage appears to be low. In such model systems as the production of hepatic tumors in rats that are given known hepatic carcinogens, one can typically produce around thousands of biochemically altered cell foci per liver, which will be followed by the appearance of several adenomas and then by one or two hepatocarcinomas (Moolgavkar et al., 1990a; Cohen and Ellwein, 1991; Luebeck et al., 1991). In examples of that sort, almost all the early lesions do not progress to cancer, but remain the same or regress; thus cancer is a rare biologic outcome. Nevertheless, the consistent association of altered cell foci with later cancer formation has prognostic value and may help in developing preventive measures.

Underlying the structural stages are molecular events, or steps, that define the beginning and end of each stage. As noted before, cancer involves a disturbance of cell growth and cell growth is under genetic regulation, so genetic damage is likely to be important in carcinogenesis.

Target genes include those related to cell division and proliferation (pro-to-oncogenes) or those which cause cells to stop dividing (anti-oncogenes or tumor-suppressor genes). The available evidence strongly supports the general concept that the cells of some cancers in humans and labora-tory animals contain activated or mutated oncogenes and, in some cancer cells, tumor-suppressor genes are inactive or missing (Weinberg, 1988; 1989). Under active investigation are the extent to which those genetic events are necessary and sufficient to result in cancer and whether the sequence of genetic events is important if more than one genetic event is necessary. Other possible target genes, such as the genes that contribute to cell-division cycles or the genes that affect the microenvironment in which developing cancer-cell clones might be inhibited or selectively enhanced, have received less attention. It is generally assumed, too, that cell proliferation in general increases the probability of inheritance of random mutations by somatic cells, thus contributing indirectly to the carcinogenic process (Cohen et al., 1991).

The concept of two-stage models emerged from Knudson's studies of heritable childhood cancers (Moolgavkar and Knudson, 1981), and was an extension of the work of Armitage and Doll (1957). For retinoblasto-ma in particular, the relation of tumor incidence to age suggested that one event is necessary in the somatic cells of hereditary carriers and two events are necessary in nonhereditary carriers. Molecular genetic analy-ses of cells from affected children have revealed that the critical event can be the loss or inactivation of both alleles of the retinoblastoma tu-mor-suppressor gene (*RB1*) (Gaillie et al., 1990). The developing retina might contain three types of cells: normal retinoblasts with two normal *RB1* alleles, intermediate retinoblasts with one altered or lost *RB1* gene, and retinoblasts with both *RB1* genes altered. In the hereditary form in which one parentally acquired allele is altered, the probability of retino-blastoma is increased, because all the developing retinoblasts have an abnormal *RB1* gene and are at risk of a second event. That three or four tumors develop in the typical gene carrier suggests that the second event is not very common. The process is limited, as a child ages, by differ-entiation of the entire embryonal retinoblast pool into adult nondividing retinal cells.

Other cancers appear more complicated. For example, mutations in both *RB1* and *p53* suppressor genes are thought to be involved, with a third presumed suppressor gene, in small-cell lung tumors (Takahashi et al., 1989). Data on colon cancer suggest five or six critical events (Vogelstein et al., 1988; Goyette et al., 1992). These examples suggest that more complex models might be required. Conceptually, the two-stage model could be extended to any number of stages.

The biologic basis of carcinogenesis is still incompletely understood. In spite of recent rapid advances at the molecular level, many of the events described cannot yet be demonstrated to be essential for the pathogenesis of cancer. Some might be incidental phenomena with no causal relationship to the carcinogenic process. Research utilizing dose-response modeling can provide insights into which events are necessary and sufficient to produce cancer by demonstrating which mechanistic assumptions are consistent with the dose-response data.

THE TWO-STAGE MODEL

The two-stage model developed by Moolgavkar, Venzon, and Knudson (Moolgavkar and Venzon, 1979; Moolgavkar and Knudson, 1981) postulates two critical events in carcinogenesis that are specific, irreversible, and hereditary (at the cell level). The model supposes three cell compartments: normal stem cells, intermediate cells that have been altered by one genetic event, and malignant cells that have been altered by two genetic events. The size of each compartment is affected by cell birth, death, and differentiation processes and by the rates of transition between cell compartments.

The model is consistent with current concepts regarding the roles of inactivated tumor-suppressor genes and activated oncogenes in carcinogenesis. It explicitly accounts for many processes considered important in carcinogenesis, including cell division, mutation, differentiation, and death and the clonal expansion of populations of cells. Although the various carcinogenic processes might have more than two steps, a major assumption is that each of them can be described as consisting of two critical, genomic events: the first is assumed to give a small growth advantage through partial abrogation of growth control, and the second is assumed to lead to total abrogation of growth control. Among the other assumptions are that a cancer arises from a single cell, that transformations of stem cells are independent events, that each transformed cell will become a tumor, and that the time required to develop from a single transformed cell into a tumor is constant.

The mathematical aspects of two-stage model development and application have been described by Moolgavkar, Cohen, and Portier and their associates (Greenfield et al., 1984; Portier, 1987; Ellwein and Cohen, 1988; Moolgavkar, 1988; Moolgavkar and Luebeck, 1990; Portier and Edler, 1990; Tan, 1991). The model permits computation of both the rate at which tumors form (incidence function) and the probability of tumor formation with respect to time. Both stochastic and deterministic forms of the model have been described.

In the schematic representation of the model (Figure 1) as described by Moolgavkar and Knudson (1981), C_0, C_1, C_2, and D represent stem cells, intermediate cells, malignant cells, and differentiated or dead cells. A normal stem cell can divide into two stem cells, die, or be transformed by mutation into an altered intermediate cell. An intermediate cell similarly divides into two intermediate cells, dies, or becomes transformed into a fully malignant cell. α_1 is the rate at which cells divide (normal cells at α_1, and intermediate cells cells at α_2), β_1 the rate at which they die, and μ_1 the rate at which they are transformed.

That formulation assumes that normal cells behave independently, which implies that they either die out (generally early in life, which would result in death of the subject) or grow exponentially throughout life. Neither alternative is realistic, so the normal stem cell population

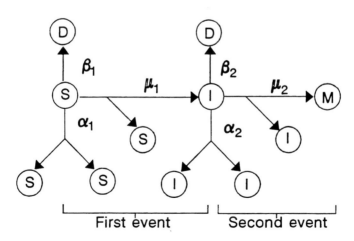

FIGURE 1   Two-stage model paradigm. Source: Moolgavkar and Knudson, 1981.

is generally modeled as growing deterministically, with intermediate cells arising from normal cells in a nonhomogeneous Poisson process with intensity function $X(s)\nu(s)$, where $X(s)$ is the (deterministic) size of the normal stem cell population at age s and $\nu(s)$ is the rate at which stem cells are converted into intermediate cells. With that form of the model, closed-form mathematical expressions can be obtained for the probability of forming a malignant cell by time t, $P(t)$, and the associated instantaneous-hazard function $h(t)$, as long as the parameters are time-independent. By integrating Expression 24 of Moolgavkar et al. (1988), one obtains

$$h(t) = \nu X \mu_2 \left(\frac{1 + a}{ab} \right) \left(\frac{1 + a}{1 + ae^{-bt}} - 1 \right) \qquad (1)$$

and

$$P(t) = 1 - \exp\left[-h(t)\right] \qquad (2)$$

where

$$h(t) = \int_0^t h(w)dw = \nu X \mu_2 \left(\frac{1 + a}{ab} \right)$$
$$\cdot \left[at + \left(\frac{1 + a}{b} \right) \mathrm{Ln} \left(\frac{1 + ae^{-bt}}{1 + a} \right) \right] \qquad (3)$$

and

$$a = \frac{1 - C_1}{C_2 - 1} \qquad (4)$$

$$b = \alpha_2 (C_2 - C_1) \tag{5}$$

$$C_1 = \frac{\beta_2 + \mu_2 + \alpha_2 - [(\alpha_2 + \beta_2 + \mu_2)^2 - 4\alpha_2\beta_2]^{1/2}}{2\alpha_2} \tag{6}$$

$$C_2 = \frac{\beta_2 + \mu_2 + \alpha_2 + [(\alpha_2 + \beta_2 + \mu_2)^2 - 4\alpha_2\beta_2]^{1/2}}{2\alpha_2} \tag{7}$$

$X(s)$ and $\mu(s)$ are not separately identifiable, and only their product can be estimated. Intermediate cells behave independently and with exponentially distributed lifespans. Consequently, clones of intermediate cells either die out or increase exponentially in size; there is no provision for growth regulation. Those assumptions are unrealistic, and alternatives have been proposed (Moolgavkar and Luebeck, 1990); however, implementation of more realistic alternatives greatly complicates the mathematical analysis and may not be necessary in providing good estimates of h(t).

The time between the occurrence of the first malignant cell and a clinically detectable cancer or death is generally modeled as a constant (Moolgavkar and Luebeck, 1990). A different assumption could easily be made, but doing so is likely to make the resulting mathematics intractable.

Before the model can be used in risk assessment, the effect of dose must be incorporated. That is generally accomplished by treating model parameters as functions of instantaneous dose (Thorslund et al., 1987; Moolgavkar and Luebeck, 1990). Dose can be incorporated into the model by introducing a dose-effect relationship into the transition rate from normal cells to intermediate cells (μ_1), into the transition rate from intermediate cells to malignant cells (μ_2), or into the growth rate of clones of intermediate cells ($\alpha_2 - \beta_2$). If the dose rate changes over time, then the corresponding parameter that dose affects is time-dependent and the solutions presented earlier do not apply. Explicit formulas for obtaining solutions when the parameters are piece-wise constant are also found in Moolgavkar and Luebeck (1990). Quinn (1989) and Moolgavkar and Luebeck (1990) showed how to obtain numerical solutions with the method of characteristics when the parameters are generally time-dependent.

One of the most important applications of dose-response models in risk assessment is to predict increased risk from exposure to low doses of a chemical. Typically, increased cancer risks on the order of $1/100$ cannot be accurately measured in either a standard animal bioassay or an epidemiologic study due to limitations of sample size, yet increased risks in human populations of this magnitude, and even smaller, are of concern.

Small increased risks from low exposures are often estimated by fitting a dose-response model to data collected at higher exposures. The form assumed for the dose-response model is of critical importance to the resulting risk estimate (NRC, 1983). Regulatory agencies have frequently applied models that assume the increased risk is linearly related to exposure (i.e., the increased risk is proportional to the amount of exposure), at least at low exposures. However, it is frequently the case that nonlinear models will fit the data equally well and predict much lower risks at low exposures. The most extreme case of a nonlinear model is a threshold model, which assumes that there is a critical exposure (i.e., a threshold) below which the risk is not increased.

As is the case with simpler descriptive models, the manner in which the effect of dose is modeled will be the determining factor in the predictions of the two-stage model at low doses. If at least one of the transition rates is assumed to vary linearly with dose at low doses and the background incidence of cancer is not zero, then the probability of cancer will vary linearly with dose at low doses, although the low dose linear slope could differ appreciably from that predicted from high dose data. However, if all the dose-related rates are assumed to vary nonlinearly with dose at low doses or to exhibit a threshold dose below which the rate is not affected by dose, then the probability of cancer will likewise vary nonlinearly with dose at low doses or exhibit a threshold below which dose cannot cause cancer, respectively. Consequently, the manner in which dose is introduced into the two-stage model is a critical assumption for risk assessment.

Depending on how the model parameters are selected, there might be a number of parameters to estimate. For example, in their application of the two-stage model to data on 1,797 rats exposed to radon, Moolgavkar and Luebeck (1990) assumed that the number of normal cells was a constant and a clone of malignant cells of any size could be identified as a tumor at necropsy. They also assumed that the rate at which interme-diate cells divided (α_2) was 10 times per week—the measured cell-division rate for adenomas in rat lungs. Despite those simplifying assumptions, six parameters still had to be estimated by fitting the model to the cancer bioassay data. Although those data constituted an extraordinarily large data set, the precision with which some of the parameters could be estimated from the tumor data was low.

Application of a six-parameter model illustrates one of the key advantages of the two-stage model, as well as one of its disadvantages. Unlike the parameters of descriptive models (e.g., statistical models not derived from underlying biologic mechanisms), the parameters of the two-stage model are required to relate to actual biologic phenomena. Thus, Moolgavkar and Luebeck made the assumption that intermediate cells are equivalent to adenoma cells. This assumption implies that adenomas progress to carcinomas and is open to investigation. Generally, the two-stage model is more useful than mainly descriptive models for testing mechanistic hypotheses of this type, because several models can be developed based on alternative biologic hypotheses, which can then be tested on the basis of goodness-of-fit. Descriptive models are developed by fitting them to data, and hypotheses regarding underlying biologic mechanisms generally cannot be tested on the basis of fit.

A disadvantage of the two-stage model illustrated by the analysis of Moolgavkar and Luebeck is the potentially large number of model parameters to be estimated. To estimate as many as six parameters reliably might require large numbers of animals exposed to various dose patterns and with serial sacrifices. Thus, most examples of application of the two-stage model have used large data sets—e.g., the ED_{01} study of 2-AAF (Cohen and Ellwein, 1990) and the study involving 1,797 animals exposed to radon (Moolgavkar and Luebeck, 1990). Such extensive data sets are available for only a few chemicals.

An alternative to estimating large numbers of parameters from tumor bioassay data is estimating specific parameters with data from other sources. Moolgavkar and Luebeck's estimation of the cell-division rate of intermediate cells on the basis of the cell-division rate in adenomas illustrates the approach. Although the approach is potentially quite useful, the knowledge needed for its general application is not yet available. In general, it requires an understanding of the steps in carcinogenesis and identification of cell types produced in the progression from normal to malignant cells, measures of the proliferation rates of intermediate cells and the rates of transformation of cells from one stage to another, and whenever these quantities are dose-related, measures of the response of the parameters to dose. The latter measures are critical in determining the dose-response relationship and consequently assessing risk and estimating potency. Ideally, one would be able to measure the dose-response relationship accurately at doses to which humans are likely to be exposed (which might be much lower than the doses that produced measurable tumor responses in a standard animal bioassay); otherwise, one must assume a functional form for the relationship, which can introduce large uncertainties.

Environmental agents, as well as interindividual genetic differences in their metabolism, can affect tumor incidence through their effects on either mutation rates or the kinetics of cell division and differentiation, or both (Nebert, 1989; 1991a,b; Nebert et al., 1991, 1993 and references reviewed therein). Numerous studies in both mice and humans have demonstrated striking genetic differences in benzo[a]pyrene-induced tumor initiation, in cigarette smoke-induced tumor initiation (and probably tumor promotion), and in dioxin-induced toxicity (and possibly tumor promotion). A mutagenic substance can increase the intermediate-cell population (be an initiator) and can also cause conversion to malignancy (be a complete carcinogen). If only the kinetics of cell division and differentiation are affected by a substance, two general outcomes would be possible: in one, cell division and differentiation would be increased equally, resulting in increased mutation rates related to the increased cell-division rate; in the second, the rate of cell division would be increased disproportionately to the rate of differentiation, and the increase would result in greater numbers of cells at risk of mutation.

APPLICATIONS OF THE
TWO-STAGE MODEL TO ANIMAL DATA

Applications of the two-stage model have been few, because of limited data availability. Standard two-year chronic carcinogenicity bioassays are not designed to provide information on the contribution of cell proliferation to tumor rates, and there are few data on the time- and dose-response effects of agents on cell proliferation. As a result, models that have been developed have generally relied on indirect measures of cell proliferation, such as increases in organ weights, or on measures of cell proliferation performed in independent experiments with protocols that provide less than ideal data for modeling purposes. Generating adequate data for the characterization of dose-response relationships for cell proliferation rates and of their contribution to tumor rates is a critical need.

The best-known examples of applications of the two-stage model to animal data are those for 2-acetylaminofluorene (2-AAF) in the mouse liver and bladder (Cohen and Ellwein, 1990), for saccharin in the rat bladder (Ellwein and Cohen, 1988), for radon in the rat lung (Moolgavkar et al., 1990b), and for *N*-nitrosomorpholine (NNM) in the rat liver (Moolgavkar et al., 1990a). In the case of 2-AAF, a genotoxic agent, liver tumor rates are consistent with its effect on the rates of transition between cell stages and have a linear dose-response relationship, as does the rate of DNA adduct formation. In the bladder, a linear rate of DNA adduct formation is observed as well, but the tumor rate is consistent with a nonlinear increase in the rate of cell proliferation at high doses and with an effect of dose on rates of transition between cell stages. Saccharin, a nongenotoxic agent, appears to induce bladder cancer as a result of toxicity-induced regenerative hyperplasia, and its dose-response model is thus based on an effect on the cell-growth rate function and not on transition rates. The model that was developed for radon is consistent with a primary effect on the rate of first transition between cell stages, a less-pronounced effect on the rate of second transition, and an increase in the proliferation rate of intermediate cells. Analysis of the data on liver-foci development associated with NNM indicates that it is a strong initiator that has a primary effect on the rate of transition to intermediate cells (as detected by foci formation) and that it has a weak promoting effect as well (as determined by its effect on the rate of foci proliferation).

The modeling approaches of Cohen and Ellwein and of Moolgavkar differ. Moolgavkar applies standard statistical methods (e.g., maximum likelihood) that have well-understood statistical properties, to a closed form solution of the two-stage model. Uncertainty in model parameters and goodness-of-fit of the model to the experimental data can be investigated using standard statistical methods. However, the procedure requires closed-form solutions for the two-stage model, which are only available for a few special cases (although more general cases can be approximated quite closely by the cases for which solutions are available). When a simulation approach is used or a closed-form solution does not exist, assessing both the model's goodness-of-fit and the uncertainty in parameter estimates is important to understand the applicability of the model for calculating risk. Sensitivity analyses that explore alternative model assumptions and parameter values should be carried out to support such an assessment. Formal statistical techniques should be used whenever possible.

On the other hand, the Cohen and Ellwein approach involves specifying values for each model parameter. A computer simulation is then used to compute realizations of the subsequent tumor response. Parameter values are varied until the realizations conform to the actual data. This procedure can be applied more generally than that of Moolgavkar because a closed-form solution is not required. However, since there are no clear criteria for selecting trial values for the parameters or for determining adequacy of fit, the most appropriate parameters may not be found and results may not be reproducible. Moreoever, the method does not readily lend itself to assessing goodness-of-fit and expressing uncertainty in parameter estimates.

Several other applications of the two-stage model have been attempted. Two examples are described here: one for the genotoxic agent benzo[a]-pyrene (B[a]P) (Clement Associates, 1988) and one for the nongenotoxic agent chlordane (Thorslund and Charnley, 1988). Both applications involve an approximation to the two-stage model that gives good results only when the probability of a tumor is not too near unity (Moolgavkar et al., 1988). The mathematical expressions defining this approximate model are much less complicated than those required for the exact model; however, the approximation will be poor in some circumstances.

These examples are presented only to illustrate particular points and do not represent complete risk analyses. The committee believes that a complete risk analysis based on a two-stage model should include statistical confidence intervals for model parameters and estimates of excess risk, which would permit determination of the ranges of risk that are consistent with the data and the particular form of the two-stage model being employed.

* *Benzo[a]pyrene.* In an inhalation study, respiratory tract tumors were induced in hamsters exposed daily to a B[a]P/sodium chloride aerosol throughout their lives. These occurred in the nasal cavity, larynx, and trachea (Thyssen et al., 1981). Doses and tumor rates are

shown in Table 1. B[a]P is assumed to increase the rates of transition between cell stages and (in the absence of information to the contrary) the B[a]P-induced rates of transition between normal and intermediate and between intermediate and cancerous cells are assumed to be equally likely. It is further assumed that these transition rates are linear functions of dose (which is likely at low doses) and that the growth rates of normal and preneoplastic cells are independent of exposure. Under these assumptions, the probability of tumor development at time t as a result of exposure to level x of B[a]P (or any other genotoxic agent for which the above assumptions are applicable) is:

$$P(x,t) = 1 - \exp\{-M(1 + Sx)^2[\exp(Gt) - 1 - Gt]/G^2\} \qquad (8)$$

where M is the background tumor rate parameter, S is the exposure-dependent transition rate between cell stages, G is the exposure-independent growth rate of intermediate cells, and t is the time (or age) at which risk is evaluated. In this case, the level x at the target tissue is assumed to be directly proportional to the administered dose.

The bioassay data in Table 1 can be used to estimate the exposure-induced relative transition rates as well as a background (spontaneous)

TABLE 1   Benzo[a]pyrene Bioassay Data: Observed and Predicted Tumor Rates

B[a]P Dose mg/m³ of Air	Average Survival Time, Weeks	Number of Hamsters with Respiratory Tract Tumors	
		Observed	Predicted
0	96.4	0/27 (0%)	0.73 (3%)
2.2	95.2	0/27 (0%)	1.88 (7%)
9.5	96.4	9/26 (35%)	9.06 (35%)
46.5	59.5	13/25 (52%)	12.59 (50%)

transition rate, using Equation 8. Fitting the equation to the data by the method of maximum likelihood and using the average survival time as the length of observation yields the following equation for determining the cancer risk for each dose level:

$$P(x,t) = 1 - \exp-[0.000115(1 + 0.312x)^2] \cdot [\exp(0.057t) - 1 - 0.057t] \qquad (9)$$

As Table 1 indicates, the number of tumors predicted by this equation is very similar to that observed. However, because none of the animals at the lowest dose developed tumors, the data are also consistent with a threshold-type dose response.

If survival had not been affected by exposure, Equation 8 could have been reduced to the simple quadratic form:

$$P(x) = 1 - \exp[-A(1 + Sx)^2] \qquad (10)$$

Equation 10 has the mathematical form of EPA's multistage model, but it is further constrained by the limitation to the first and second powers of x, and it has only two free parameters to determine the three coefficients in the exponent.

If 96.4 weeks is used as the average survival time for the control (unexposed) group, the following time-independent lifetime-risk relationship is obtained with the simplified form of the equation:

$$P(x) = 1 - \exp[-0.0272(1 + 0.312x)^2] \qquad (11)$$

With this simplified form and an assumption that at low doses the product of the linear term of the equation and the dose is a close approximation of the estimated cancer risk, the linear term may be expressed as

$$q_1 = (2)(0.0272)(0.312) = 0.0170 \, (mg/m^3)^{-1}. \qquad (12)$$

To extrapolate that potency value to a human cancer risk estimate for lung cancer and B[a]P, the experimental exposure periods (4.5 hours/day, 7 days/week, for 10 weeks, then 3 hours/day thereafter) can be converted to an average 24-hour exposure period to yield a human cancer potency value of

$$q_1 = (0.0170)(24/3.147) = 0.1295 \, (mg/m^3)^{-1}. \qquad (13)$$

That value can be compared with the cancer potency value of 1.7 $(mg/m^3)^{-1}$ calculated using the linearized multistage modeling procedure by EPA (1980) based on the same bioassay data. This approximately order-of-magnitude difference is due, at least in part, to the use of the 95% upper confidence limit by EPA instead of the maximum likelihood estimate.

• *Chlordane.* The termiticide chlordane produces liver tumors in CD-1 mice (IRDC, 1973), promotes the incidence of liver tumors initiated by dimethylnitrosamine in B6C3F$_1$ mice (Williams and Numoto, 1984), and is only a weak mutagen (Cavender et al., 1986). As a result, its tumorigenicity has been attributed to its tumor-promoting, not-initiating, ability. A nongenotoxic mechanism of action may be proposed that involves hepatocellular mitogenesis, on the basis of observation of proliferative activity in the livers of exposed animals (IRDC, 1973). In the context of the two-stage model, chlordane's tumorigenicity could be proposed to result from its ability to increase the birth rate of intermediate cells while having no direct effect on the rates of transition between cell stages. The probability that a tumor will develop by time t after constant exposure to chlordane at dose x under those assumptions can be expressed as

$$P(x,t) = 1 - \exp - \{M(\exp G[x]t - 1 - G[x]t)/G^2(x)\}, \qquad (14)$$

where $M = M_0M_1$ and, again, the approximate form of the two-stage model has been applied. Using this model, the growth rate of intermediate cells, $G(x)$, may be expressed in the form

$$G(x) = G(0) + (G[\infty] - G[0])R(x), \qquad (15)$$

where G(0) is the normal intermediate-cell growth rate, G(∞)is the upper
bound on the chlordane-induced intermediate-cell growth rate, and R(x)
is the fraction of the maximal increase in the intermediate-cell growth
rate that is induced by a constant exposure to chlordane at dose x.

The functional form chosen for R(x) was that of a bounded log-logistic
function on the basis of the following logic: at low doses, the function
is bounded by the background number of intermediate cells; at high
doses, it is reasonable to assume that proliferation, and therefore the
number of intermediate cells, reaches a plateau, because it cannot increase indefinitely. That logic has not been examined carefully or validated because of the difficulty inherent in identifying intermediate cells
and their kinetics. The log-logistic form assumed for the growth-rate
function is

$$R(x) = (1 + \exp -[I + S\ln x])^{-1}, \qquad (16)$$

where chlordane is assumed to elicit its proliferative effect as a result of
binding at a cellular receptor of some kind, I is proportional to its binding constant, and S is the average number of receptors in (or on) the
affected cells.

TABLE 2   Chlordane Bioassay Data: Observed and Predicted Tumor Rates

| | Number of Mice with Liver Tumors | | | |
| | Male | | Female | |
Dose, ppm	Observed	Predicted	Observed	Predicted
0	3/33 (9%)	3.0 (9%)	0/45 (0%)	2.5 (6%)
5	5/55 (9%)	5.8 (11%)	0/61 (0%)	3.9 (6%)
25	41/52 (79%)	41.0 (79%)	32/50 (64%)	30.4 (61%)
50	32/39 (82%)	32.0 (82%)	26/37 (70%)	23.8 (64%)

Source: Thorslund and Charnley, 1988

Equation 14 can be fitted to the bioassay data for chlordane shown in Table 2 by using the log-logistic form of the growth-rate function described above and making a number of assumptions regarding the needed parameter estimates (see Table 3):

- Use human age-specific liver cancer death rates to estimate $G(0)$.
- Use the background rate of spontaneous liver tumors in CD-1 mice to estimate M in male mice and the maximum-likelihood method and total tumor response to estimate M in female mice.
- Fix S at 4 (based on analogy to other logistic responses).
- Estimate $G(\infty)$ and I by equating the observed tumor rates at the two highest chlordane doses to the parametric form of the model and solving the resulting two nonlinear equations with two unknowns.

TABLE 3 Parameter Estimates for Chlordane Dose-Response Model

Parameter	Estimate
$G(0)$	0.06314
M (males)	2.8622×10^{-6}
M (females)	1.722×10^{-6}
$G(\infty)$	0.11527
I	-9.524
S	4.0

Source: Thorslund and Charnley, 1988.

The predicted tumor rates are quite similar to the observed rates (Table 2) using Equation 16 and the parameter estimates specified in Table 3. The human cancer dose-response model developed on the basis of mouse data for chlordane yields the estimates of cancer risk shown in Table 4. These predictions are at least 3 orders of magnitude less than those obtained using the EPA's linearized multistage procedure, also shown.

To further explore the utility of two-stage models in risk assessment, the committee has conducted additional calculations involving the chlordane example. In the chlordane analysis conducted by Thorslund and Charnley, it was assumed that the two-stage model is appropriate for chlordane. It was further assumed that chlordane increases the rate of division of intermediate cells, but otherwise does not affect tumor rates. The latter is a critical assumption, because, even if the two-stage framework is appropriate for chlordane, the weak dose-response relationship

TABLE 4   Estimates of Chlordane's Human Cancer Risk

	Lifetime Cancer Risk	
Dose, ppm	Tumor Promotion Model	Linearized Multistage Model
1	1.6×10^{-5}	4.8×10^{-2}
0.1	1.2×10^{-9}	4.9×10^{-3}
0.01	8.8×10^{-14}	4.9×10^{-4}

Source: Thorslund and Charnley, 1988

of chlordane could be considerably different from that predicted by the model if chlordane also affects some other step in the carcinogenesis process (e.g., if chlordane also had some effect on the rate of transition from normal to intermediate cells).

Even if all those assumptions are correct, the critical assumption regarding how chlordane affects the growth rate of intermediate cells still requires validation. Having no data for determining that growth rate, Thorslund and Charnley assumed that it had a particular mathematical form, expressed as $G(x)$, where x is the dose of chlordane. To explore the sensitivity of the risk-assessment results to this assumption, the committee experimented with other functional forms to determine whether other forms might also describe the data but yield different risk-assessment results. One functional form considered by the committee, which varies only slightly from that of Thorslund and Charnley but that also describes the data was

$$G(x) = 0.063088 + 0.052182 \cdot \{1 - \exp[-0.001(1 + 0.56x)^{-3}\}. \tag{17}$$

That expression was substituted for the formula for cell-proliferation rate assumed by Thorslund and Charnley, which was

$$G(x) = 0.06314 + 0.05213\{1 + \exp[9.524 - 4\ln(x)]\}^{-1}. \tag{18}$$

All other parameter values used by Thorslund and Charnley were retained (M = 2.8622 x 10^{-6}, t = 78 weeks).

Table 5 shows that the fit of Equation 17 to the chlordane data on male mice is comparable with that of the model used by Thorslund and Charnley. However, as Table 6 indicates, the predictions of the two models differ sharply at doses below the experimental range, which is generally the range of interest for environmental regulation. The risk estimates calculated with the committee's exploratory model are higher than those obtained with the model of Thorslund and Charnley by a factor of about 39 for a chlordane dose of 1 ppm, about 24,000 for a dose of 0.1 ppm, and about 23,000,000 for a dose of 0.01 ppm. Those large differences are due solely to differences in the assumed cell-proliferation rate G(x). Thus, distinguishing between these risk estimates would require distinguishing between the underlying cell-proliferation rates determined by the two alternative expressions for G(x).

TABLE 5   Chlordane Data: Fit of Alternative Approximate Two-Stage Models

| | Number Male Mice with Liver Tumors | | |
| | | Predicted | |
Dose, ppm	Observed	Thorslund and Charnley Model[a]	Alternative Model[b]
0	3/33 (9%)	3.0 (9%)	3.0 (9%)
5	5/55 (9%)	5.5 (9%)	5.6 (10%)
25	41/52 (79%)	41.0 (79%)	41.0 (79%)
50	32/39 (82%)	32.0 (82%)	32.1 (82%)

[a] based on Expression (18) for G(x).
[b] based on Expression (17) for G(x).

Table 6 compares the values of the cell-proliferation rates from the

Thorslund and Charnley model with those from the alternative model at both the experimental doses and the lower doses for which additional risk was estimated. The predictions of cell-proliferation rates agree closely both at the experimental doses and at lower doses. Extremely small differences in cell-proliferation rate can result in large differences in additional risk. At doses of 0.1 and 0.01 ppm, the two cell-proliferation rates differ only in the sixth decimal place, whereas the resulting extra risks differ by factors of about 4 and 7 orders of magnitude, respectively. Thus, tiny changes in the cell-proliferation rate can make enormous differences in the resulting risk estimates. Given the variation that is normal in biologic systems, it is highly unlikely that such small differences in cell-proliferation rate could ever be accurately distinguished. An additional source of uncertainty could be introduced by assuming that as a weak mutagen, chlordane could have an effect on the transition rates in addition to the cell growth rates. This assumption could alter the risk estimates even more.

TABLE 6   Low-Dose Cancer Risk Estimates for Chlordane Derived from Two-Stage Models[a]

Dose, ppm	Additional Risks		Cell-proliferation Rate, $G(x)$	
	Thorslund and Charnley Model	Alternative Model	Thorslund and Charnley Model	Alternative Model
1.0	1.6×10^{-5}	3.7×10^{-4}	0.063144	0.063286
0.1	1.6×10^{-9}	3.9×10^{-5}	0.063140	0.063149
0.01	1.6×10^{-13}	6.2×10^{-6}	0.063140	0.063141
0	--	--	0.063140	0.063140

[a]Based on data in Table 2 on liver tumors in male mice.

In the committee's discussions of these results, it was suggested that they might be due to the use of the approximation to the two-stage model and that the exact form of the model might not exhibit such instabilities. To explore that issue, the exact form of the two-stage model was fitted to the chlordane data on male rats. As in the application of the approximate model, chlordane was assumed to affect the division and death of intermediate cells. The following specific parameter values were used in the fitting:

$$X = 10^7 \text{ cells}$$
$$\nu = 2.7 \times 10^{-7}$$
$$\mu_2 = 1 \times 10^{-6}$$
$$\alpha_2 = G(x)/0.4949$$
$$\beta_2 = G(x)$$

As before, x is the dose of chlordane. The two specific forms of $G(x)$ (Expressions 17 and 18) applied to the approximate solution were also applied here in connection with the exact solution. The resulting exact solutions are virtually indistinguishable from the corresponding approximate solutions. Table 7 shows that both exact models fit the data on male mice almost exactly, just as the approximate models do. Table 8

TABLE 7   Chlordane Data: Fit of Alternative Exact Two-Stage Models

	Number Male Mice with Liver Tumors		
		Predicted	
Dose, ppm	Observed	Expression 18	Expression 17
0	3/33 (9%)	3.0 (9%)	3.0 (9%)
5	5/55 (9%)	5.5 (10%)	5.7 (10%)
25	41/52 (79%)	41.2 (79%)	41.2 (79%)
50	32/39 (82%)	32.1 (82%)	32.1 (82%)

shows virtually the same risks at low doses for the exact models as shown in Table 6 for the approximate models.

The two exact and two approximate dose responses are depicted graphically in Figures 2 and 3. Figure 2 shows that the exact expression for the probability of response, $P(x)$, agrees closely with the approximate solution throughout the complete range of exposures when both are based on the same expression for $G(x)$. It also shows that the two expressions for $G(x)$ provide comparable response probabilities at the experimental exposure levels (0 ppm, 5 ppm, 25 ppm, and 50 ppm) and at all exposure levels below 5 ppm. Figure 3, which is the same as Figure 2 except that log scales are used and the vertical axis is the additional probability of response induced by exposure $[P(x) - P(0)]$, shows that the exact solution for additional probability also agrees closely with the corresponding approximate solution over a wide exposure range, including very low exposures. It also shows that the two expressions for $G(x)$ provide similar results for additional risk at high exposures but very different values at low exposures.

TABLE 8 Low-Dose Cancer Risk Estimates for Chlordane Derived from Exact Two-Stage Models

| Dose, ppm | Additional Risks | |
	Expression 18 for $G(x)$	Expression 17 for $G(\times)$
1.0	1.7×10^{-5}	6.4×10^{-4}
0.1	1.7×10^{-9}	4.0×10^{-5}
0.01	1.7×10^{-13}	3.8×10^{-6}

Thus, the exact models produce results in this case that are virtually indistinguishable from those produced by the approximate models that use the same cell-proliferation rate function $G(x)$. The two expressions for $G(x)$ provide very similar response probabilities at the experimental exposure levels (and therefore are indistinguishable based on the experimental data) but predict divergent estimates of additional risk at low exposures. Consequently, there is virtually no difference between the exact and approximate two-stage solutions in this case, and the exact model is subject to the same instabilities as the approximate model.

This type of instability is likely to be the rule, rather than the exception. A general model for the probability of cancer arising from a dose x can be written as

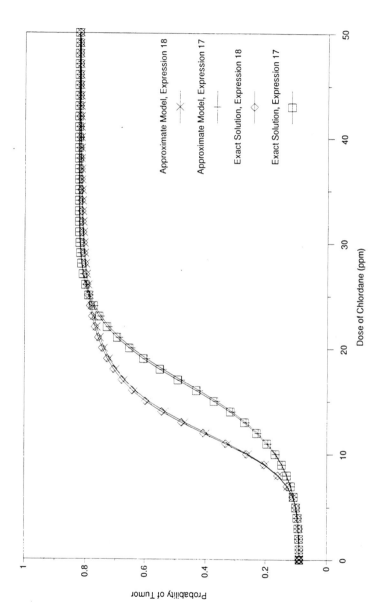

FIGURE 2 Comparison of exact and approximate two-stage models applied to chlordane data on male mice.

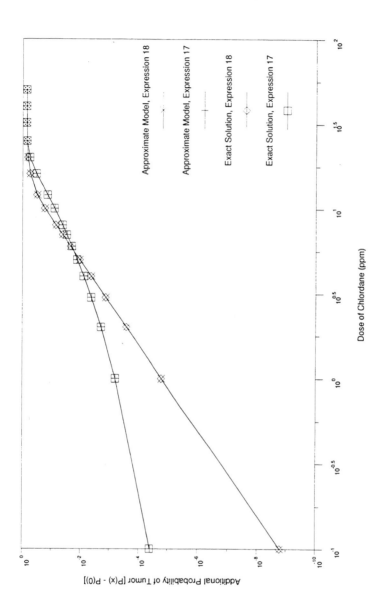

FIGURE 3 Comparison of exact and approximate two-stage models at low exposures applied to chlordane data on male mice.

$$P(x) = H[\beta_1, \ldots, \beta_k, G(x)], \tag{19}$$

where $\beta_1, \ldots, \beta_k$ are parameters that are unaffected by dose and $G(x)$ is the parameter that is affected by dose. In the chlordane example, $G(x)$ represented cell-proliferation rate, although for the purposes of the current argument it could be any biologic parameter that is affected by dose. At low doses, the probability is approximately

$$P(x) = P(0) + \left[\frac{\partial P}{\partial G} \Big|_{x=0} \right] [G(x) - G(0)]. \tag{20}$$

Thus, the dose-response relationship will behave at low doses like a linear function of $G(x)$. That implies that the dose-response relationship for cancer at low doses will mimic that of the parameter that dose affects. The Red Book (NRC, 1983) showed that different dose-response curves for $P(x)$ could be obtained that fit data in the observable range but yield results for incremental risk (above background) that differ by many orders of magnitude in the low-dose range. The same arguments apply to $G(x)$ and therefore, through Equation 20 relating $P(x)$ to $G(x)$, to $P(x)$ again.

Those considerations suggest steps that are critical in using a two-stage model (or any other biologically based model) for low-dose extrapolation. Identifying the biologic steps that lead to cancer and determining which ones are affected by the chemical insult are the first steps in the process. Another is the specification of the dose-response relationship for the parameter $G(x)$ (cell-proliferation rate, mutation rate, etc.) or parameters that are affected by the chemical insult. Regardless of how detailed and reliable the model is otherwise, if it does not specify a mathematical form for $G(x)$, the quantitative predictions of the model at low doses are essentially arbitrary. More precisely, given a specific two-stage model that fits a given set of dose-response data adequately, the function G can be adjusted so that the adjusted model describes the data equally well but corresponds to estimates of additional cancer risk over the background risk that differ from estimates based on the original model by arbitrarily large factors at any dose below the lowest experimental dose.

Prescribing with confidence the mathematical form of the dose-response relationship for any particular biologic parameter that depends on dose is likely to be difficult. Kopp and Portier (1989) found that when the approximate form of the model fails to characterize accurately the cumulative distribution function of the time to tumor onset, bias may result in the estimates of the remaining parameters. The critical need in applying biologically based models will be for data on the response of the model parameters that are affected by dose. While the committee encourages the use of formal statistical methods, the application of such methods to estimate model parameters from bioassay data does not resolve uncertainty about the relationship of these parameters to dose at low doses. Bioassay data such as those for benzo[a]pyrene in Table 1 and chlordane in Table 2 do not provide a basis for determining the shape of a dose-response relationship at low doses. These data sets would be consistent with a model that predicted zero incremental cancer risk at the lowest positive dose. These data sets are also consistent with the predictions of incremental risk of 2% to 6% over background at the lowest positive dose, as shown in Tables 1 and 2. The shape of the dose-response relationship will be determined by assumptions about how the parameters in the model depend upon dose, supplemented by direct measurements of cell kinetics to the extent that such measurements are available. As the chlordane example illustrates, alternative functional forms that fit the data in the experimental range can lead to widely differing estimates of risk in the low-dose range. Narrowing the uncertainty in the low-dose range will require improved mechanistic understanding of how exposure to low doses of a toxicant affects the kinetics of cell transformation and proliferation.

DISCUSSION

Data Needs

The strength of the two-stage model is its ability to use information about cell division and differentiation. However, many of the discrete steps in those processes cannot be well characterized.

Among the problems is our relative inability to identify the cells in the several compartments. For many tissues, the stem cell populations are still unknown or structurally indistinguishable from related cell populations. It is probably not correct to assume that all cells that can divide or form adducts are necessarily at risk of transformation. We need biologic markers to identify the susceptible cell populations. The intermediate cell populations are also often difficult to identify. The many putative preneoplastic lesions associated with the carcinogenic process include few for which a causal association has been demonstrated. Malignant cells themselves might be difficult to identify until a tumor clone has grown enough for histopathologic diagnosis.

Measuring birth and death processes and transition rates requires identification of the cells in the several compartments. Birth processes are relatively easy to measure with existing methods, but methods for measuring programmed and unprogrammed cell death are still under development. In addition, the intermediate cell clones themselves are not always homogeneous, and cells can differ considerably from one another in biologic potential.

Those considerations and many others (including the doses to the target cells and interindividual differences in chemical metabolism) apply not only to laboratory animals, but also to humans, for whom similar information on cell kinetics is required. Humans pose the additional complication of greater heterogeneity (genetic and environmental) with individual variability in susceptibility to tumor formation at various body sites. To assess risk, one needs information not only about processes that take place in unexposed subjects, but also about the effect of various doses on the processes themselves. That is true, regardless of the dose-response modeling procedures used.

Criteria for Adoption

Before the two-stage model can be adopted for routine health risk assessments, chronic bioassay methods will have to be changed to generate the necessary data. It will be helpful, too, to evaluate the methods through a series of studies that use various agents in multiple animal strains or species. Prospective hypothesis-testing studies are preferable to retrospective model-fitting exercises.

The two-stage model can be used now to gain insights into the nature of induced carcinogenesis. The examples discussed illustrate the usefulness of two-stage models in characterizing the critical events. They also reveal the types and numbers of assumptions that might be made when data are incomplete or lacking. The models could be used as well to examine a range of assumptions.

The committee encourages diverse applications of the two-stage model to gain insight into its usefulness, particularly for risk assessment. However, the committee also recommends that, whenever the model is applied in formal risk assessment or hypothesis-testing, the reproducibility and scientific validity of the results be ensured by the application of optimal statistical methods (e.g., maximum-likelihood methods) to estimate values of parameters and to test goodness-of-fit. Critical assumptions (those with a major quantitative impact on risk estimates) should be clearly stated. And statistical confidence-interval methods, sensitivity analyses, and related quantitative methods should be applied as appropriate to determine the extent to which the resulting data are consistent with other mathematical representations and ranges of risk.

For the time being, the committee recommends that two-stage models be used primarily to promote research understanding; for health risk assessments, two-stage models can be used in conjunction with other models to add perspective to the evaluation process.

Prospects

Until recently, information about the stages of carcinogenesis has been largely limited to the descriptive and operational terms of "early" and "late" effects in epidemiologic studies and "initiation" and "promotion" in animal studies. The current growth of concepts and information about molecular carcinogenesis in patients and in experimental systems, however, promises new opportunities for conceptual understanding and model development. As more mechanistic information becomes available, the results of some human and animal studies can be expected to converge and make extrapolations across species more precise. Moreover, the patterns of genetic alterations in preneoplastic and neoplastic cells will probably help to distinguish tumors induced by exposure to specific environmental agents from those in the background (of endogenous or unknown origin) and so lead to better measures of attributable risks. Biologically based mathematical models will continue to evolve in concert with advances in biology and medicine.

CONCLUSIONS AND RECOMMENDATIONS

Mechanistic understanding of toxicity has strong implications for improvement in the development of low-dose extrapolation for the regulation of chemical substances. Currently, low-dose extrapolation uses a multistage model with data developed from human occupational exposures or from whole animal bioassays (Anderson et al., 1983). The newer two-stage model examined in this report attempts to use data related more to mechanisms of toxicity. Its potential utility (and the gathering of the data needed to use it) derives from the recent rapid development of biologic investigative techniques.

Understanding of and data on cell birth and death are required for the development and use of the two-stage model, but they do not exist for most chemicals. A better mechanistic understanding must be developed, if modeling efforts are to take advantage of cell birth and death data. If the mechanism of a toxic effect is not understood, inappropriate dose-response data are likely to be used in the extrapolation process, which could then produce an incorrect result. The committee recommends that when critical assumptions about mechanisms of toxicity are made, they must be clearly stated.

The two-stage model can be used as a basis for decision-making, if there is sufficient mechanistic understanding and if a sufficient data base is available. The committee recognizes that regulators face the question of how to determine when such understanding, data, and models are sufficient and appropriate; no hard and fast rules can be given. Complicating the issue is that there is a continuum in the extent of mechanistic understanding and data on any chemical. The risk-management context need also be considered.

Scientific work on the two-stage model of carcinogenesis has proceeded sufficiently for it to be clear that its further development should be strongly encouraged. Regulatory agencies might review decisions and standards on materials with economic or public-health importance to see whether enough data are available or can be rapidly collected to permit application of the two-stage model for additional perspective. The judgment of scientists as to whether sufficient data are available or could be collected might be helpful to regulatory agencies before they decide to apply this newer model in risk assessment. Experience in conducting such reviews will probably lead to a set of criteria for determining when the two-stage model should be used. The proposed reviews should be conducted on only a narrowly limited number of materials. And they must not be allowed to substitute for or interfere with the prompt regulation of or setting of standards for materials currently or soon to be under examination.

The committee recommends exploratory applications of the two-stage model along with its testing and validation. A first stage in the testing requires mechanistic understanding and the gathering of sufficient data to permit its use. Comparative information on humans must be developed as a part of the validation process. The committee also recommends that statistical confidence-interval methods, sensitivity analyses and related quantitative methods as appropriate be applied to determine the extent that the data are consistent with other mathematical representations and ranges of risk. Although the committee recognizes that the simulation approach to model fitting can have very important uses, particularly for exploratory data analysis and when no closed form solution of the two stage model is available, the committee recommends that whenever the model is applied in formal risk assessment, formal statistical methods (e.g., maximum likelihood) should be employed.

References

Anderson, E.L. and the Carcinogen Assessment Group of the U.S. Environmental Protection Agency. 1983. Quantitative approaches in use to assess cancer risk. Risk Anal. 3:277-295.

Armitage, P., and R. Doll. 1957. A two-stage theory of carcinogenesis in relation to the age distribution of human cancer. Br. J. Cancer 11:161-169.

Balmain, A., and K. Brown. 1988. Oncogene activation in chemical carcinogenesis. Adv. Cancer Res. 51:147-182.

Barbacid, M. 1987. ras genes. Ann. Rev. Biochem. 56:779-827.

Cavender, F.L., B.T. Cook, N.P. Page, V.J. Cogliano, and A.M. Koppibar. 1986. Carcinogenicity Assessment of Chlordane and Heptachlor/heptachlor Epoxide. EPA 600/6-87/004. U.S. Environmental Protection Agency, Research Triangle Park, N.C.

Clement Associates, Inc. 1988. Comparative Potency Approach for Estimating the Cancer Risk Associated with Exposure to Mixtures of Polycyclic Aromatic Hydrocarbons. Interim Final Report. Contract #68-02-4403, prepared for the U.S. Environmental Protection Agency, Research Triangle Park, N.C.

Cohen, S.M., and L.B. Ellwein. 1990. Proliferative and genotoxic cellular effects in 2-acetylaminofluorene bladder and liver carcinogenesis: Biological modeling of the EDO1 study. Toxicol. Appl. Pharmacol. 104:79-93.

Cohen, S.M., and L.B. Ellwein. 1991. Genetic errors, cell proliferation, and carcinogenesis. Cancer Res. 51:6493-6505.

Cohen, S.M., D.T. Purtilo, and L.B. Ellwein. 1991. Pivotal role of increased cell proliferation in human carcinogenesis. Mod. Pathol. 4:371-382.

Ellwein, L.B., and S.M. Cohen. 1988. A cellular dynamics model of experimental bladder cancer: Analysis of the effect of sodium saccharin in the rat. Risk Anal. 8:215-221.

EPA (U.S. Environmental Protection Agency). 1980. Ambient Water Quality Criteria for Polynuclear Aromatic Hydrocarbons. EPA 440/5-80-069. Environmental Criteria and Assessment Office, Cincinnati, Ohio, and the Office of Water Regulations and Standards, Washington, D.C. Available as NTIS 81-117806.

Gaillie, B.L., J.A. Squire, A. Goddard, J.M. Dunn, M. Canton, D. Hinton, X. Zhu, and R.A. Phillips. 1990. A biology of disease: Mechanisms of oncogenesis in retinoblastoma. Lab. Invest. 62:394-408.

Goyette, M.C., K. Cho, C.L. Fasching, D.B. Levy, K.W. Kinzler, C. Paraskeva, B. Vogelstein, and E.J. Stanbridge. 1992. Progression of colorectal cancer is associated with multiple tumor suppressor gene defects but inhibition of tumorigenicity is accomplished by correction of any single defect via chromosome transfer. Mol. Cell Biol. 12:1387-1395.

Greenfield, R.E., L.B. Ellwein, and S.M. Cohen. 1984. A general probabilistic model of carcinogenesis: Analysis of experimental urinary bladder cancer. Carcinogenesis 5(4):437-445.

Hollstein, M., D. Sidransky, B. Vogelstein, and C.C. Harris. 1991. p53 mutations in human cancers. Science 253:49-53.

IRDC (International Research and Development Corporation). 1973. Chlordane: Eighteen-month oral carcinogenesis: Analysis of experimental urinary bladder cancer. Report to Velsicol Corporation.

Kopp, A., and C.J. Portier. 1989. A note on approximating the cumulative distribution function of the time to tumor onset in multistage models. Biometr. 45:1259-1263.

Luebeck, E.G., S.H. Moolgavkar, A. Buchman, and M. Schwarz. 1991. Effects of polychlorinated biphenyls in rat liver: Quantitative analysis of enzyme-altered foci. Toxicol. Appl. Pharmacol. 111:469-484.

Mantel, N., and W.R. Bryan. 1961. "Safety" testing of carcinogenic agents. J. Natl. Cancer Inst. 27:455-470.

Moolgavkar, S.H. 1988. Biologically motivated two-stage model for cancer risk assessment. Toxicol. Lett. 43(1-3):139-150.

Moolgavkar, S.H., and D.J. Venzon. 1979. Two-event models for carcinogenesis: Incidence curves for childhood and adult tumors. Math. Biosci. 47:55-77.

Moolgavkar, S.H., and A.G. Knudson. 1981. Mutation and cancer: A model for human carcinogenesis. J. Natl. Cancer Inst. 66(6):1037-1052.

Moolgavkar, S.H., and G. Luebeck. 1990. Two-event model for carcinogenesis: Biological, mathematical, and statistical considerations. Risk Anal. 10:323-341.

Moolgavkar, S.H., A. Dewanji, and D.J. Venson. 1988. A stochastic two-stage model for cancer risk assessment. I. The hazard function and the probability of tumor. Risk. Anal. 8:383-392.

Moolgavkar, S.H., E.G. Luebeck, M. de Gunst, R.E. Port, and M. Schwarz. 1990a. Quantitative analysis of enzyme-altered foci in rat hepatocarcinogenesis experiments I: Single agent regimen. Carcinogenesis 11:1271-1278.

Moolgavkar, S.H., F.T. Cross, G. Luebeck, and G.E. Dagle. 1990b. A two-mutation model for radon-induced lung tumors in rats. Radiat. Res. 121:28-37.

NRC (National Research Council). 1983. Risk Assessment in the Federal Government. Washington, D.C.: National Academy Press. 191 pp.

Nebert, D.W. 1989. The *Ah* locus: Genetic differences in toxicity, cancer, mutation, and birth defects. Crit. Rev. Toxicol. 20:153-174.

Nebert, D.W. 1991a. Polymorphism of human *CYP2D* genes involved in drug metabolism: Possible relationship to interindividual cancer risk. Cancer Cells 3:93-96.

Nebert, D.W. 1991b. Identification of genetic differences in drug metabolism: Prediction of individual risk of toxicity or cancer. Hepatol. 14:398-401.

Nebert, D.W., D.D. Petersen, and A. Puga. 1991. Human *Ah* locus polymorphism and cancer: Inducibility of *CYP1A1* and other genes

by combusion products and dioxin. Pharmacogen. 1:68-78.
Nebert, D.W., A. Puga, and V. Vasiliou. 1993. Role of the *Ah* receptor and the dioxin-inducible [*Ah*] gene battery in toxicity, cancer and in signal transduction. Ann. N.Y. Acad. Sci. (In press)
Portier, C.J. 1987. Statistical properties of a two-stage model of carcinogenesis. Environ. Health Perspect. 76:125-131.
Portier, C.J., and L. Edler. 1990. Two-stage models of carcinogenesis, classification of agents, and design of experiments. Fund. Appl. Toxicol. 14:444-460.
Quinn, D.W. 1989. Calculating the hazard function and probability of tumor for cancer risk assessment when the parameters are time-dependent. Risk Anal. 9:407-413.
Takahashi, T., M.M. Nau, I. Chiba, M.J. Birrer, R.K. Rosenberg, M. Vinocour, M. Levitt, H. Pass, A.F. Gazdar, and J.D. Minna. 1989. p53: A frequent target for genetic abnormalities in lung cancer. Science 246:491-494.
Tan, W.-Y. 1991. Stochastic Models of Carcinogenesis. New York: Marcel Dekker, Inc. 249 pp.
Thorslund, T.W., and G. Charnley. 1988. Quantitative dose-response models for tumor promoting agents. Pp. 245-256 in Carcinogen Risk Assessment: New Directions in Qualitative and Quantitative Aspects, R.W. Hart and F.D. Hoerger, eds. Banbury Report #31. Cold Spring Harbor Laboratory, Cold Spring Harbor, N.Y.
Thorslund, T.W., C.C. Brown, and G. Charnley. 1987. Biologically motivated cancer risk models. Risk Anal. 7:109-119.
Thyssen, J., J. Althoff, G. Kimmerle, and U. Mohr. 1981. Inhalation studies with benzo[a]pyrene in Syrian golden hamsters. J. Natl. Cancer Inst. 66:575-577.
Williams, G.M., and S. Numoto. 1984. Promotion of mouse livewr neoplasms by the oranochlorine pesticides chlordane and heptachlor in comparison to dichlorordiphenyltrichloroethane (DDT). Carcinogenesis 5:1689-1696.
Vogelstein, B., E.R. Fearon, S.R. Hamilton, S.E. Kern, A.C. Preisinger, M. Leppert, Y. Nakamura, R. White, A.M.M. Smits, and J.L. Bos. 1988. Genetic alterations during colorectal-tumor development. N.E. J. Med. 319:525-532.
Weinberg, R.A. 1988. The genetic origin of human cancer. Cancer

61:1963-1968.

Weinberg, R.A. 1989. Oncogenes, antioncogenes and the molecular bases of multistep carcinogenesis. Cancer Res. 49:3713-3721.

Williams, G.M., and S. Numoto. 1984. Promotion of mouse liver neoplasms by the organochlorine pesticides chlordane and heptachlor in comparison to dichlorodiphenyltrichloroethane (DDT). Carcinogenesis 5:1689-1696.

Appendix A

Workshop Summary

TWO-STAGE MODEL OF CARCINOGENESIS

The goals of this workshop were (1) to assess the scientific basis for the two-stage model of carcinogenesis and (2) to evaluate the possible applications of the two-stage model to the health risk assessment process. Two-stage models are based on the assumptions that carcinogenesis is a multistage process, and that in its simplest form, two critical events are sufficient to convert normal cells to cancer cells (e.g., retinoblastoma in children).

The workshop was opened by the vice-chair, D. Mattison, who welcomed the participants and provided perspective on the relation of this workshop to the overall activities of the Committee on Risk Assessment Methodology (CRAM). The workshop chair, R. Griesemer, emphasized that the workshop is one mechanism through which CRAM obtains information and urged the participants to share additional ideas or information with CRAM after the workshop.

BIOLOGICAL FACTORS IN TWO-STAGE MODELS

A.J. Knudson, who first proposed the concept of two-stage models, presented a keynote address on the evidence from studies of heritable cancers in humans that supports the concept of two-stage models.

If cancer is related to somatic mutations, there should be some background incidence for all cancers. One would anticipate that there would be an increase in incidence upon exposure to agents that affect this process and that there would be specific targets for those mutations with tissue specificity.

At present we know of two classes of targets, proto-oncogenes and anti-oncogenes (suppressor genes). Where oncogene mutations are found in human tumors, the evidence indicates they may not be the initial events; in some instances, specific translocations seem to be the only identifiable event in the origin of a cancer. The translocations seem to be dominant in the sense that activation of one copy of an oncogene confers malignancy on a cell. In the case of suppressor genes, with release of control of cell growth, two copies must become inactivated and the events can be hereditary or nonhereditary.

Hereditary cancers have provided useful information about the genetic events in carcinogenesis. Virtually every cancer type has a dominantly inherited subgroup. The hereditary fraction for retinoblastoma is rather large (about 40%). The probability of hereditary retinoblastoma in children with an inherited abnormal *rb* gene is 100,000 times greater than that for the nonhereditary form. The relation of incidence to age suggests that one event is necessary in somatic cells of carriers and two events in nonhereditary cases. Now that the gene has been isolated and mapped on chromosome 13, this suggestion has been supported by genetic analyses of cells from affected children. Somatic mutations depend on the mutation rate per cell division and the number of cell divisions per unit time; retinoblastomas do not develop in adults because the retinal cells have differentiated and no longer divide.

In embryonal tissues such as retina, there is no conditional cell division; a mutation results in a clone of cells carrying the mutation. One can imagine initiation as a loss of one retinoblastoma suppressor gene and promotion as the proliferation that normally occurs in retinoblasts.

Survivors of the hereditary form of retinoblastoma are at risk for other cancers. About 15% of gene carriers develop osteosarcoma. About 95% of patients with osteosarcoma have mutations in both the *rb* and *p*53 genes. These two genes are involved in virtually all small cell lung tumors but a third presumed gene on chromosome 3*p* is also involved in 90% of those tumor cases. A comparison of these three tumors that appear to involve a different number of genes (one for retinoblastoma, two for osteosarcoma, three for small cell lung tumors) suggests that as the number of genes involved goes up, the relative risk goes down because the inherited gene is a smaller and smaller fraction of the total number of important events.

For adult carcinomas, the picture is less clear. Renewal tissues in which stem cells replicate may have some properties like embryonal tissues and a number of gene changes are being found in a variety of human tumor types (e.g., colon). Still unknown, however, is how many genetic events are required for a particular cancer and what is the meaning of the various events.

R. Maronpot's discussion dealt with oncogenes and cell proliferation from the perspective of an experimentalist. He suggested that the one-hit model may be more appropriate for cancers that arise from exposure to ionizing radiation, potent alkylating agents, or in transgenic mice developing lung cancer, for example, but he admitted that the multi-clonal nature of the response indicated that a second event would have to be postulated. The mouse skin tumor model is a wellknown example of two-stage tumors induced by xenobiotics. He cautioned that the data to which models may be applied may have implied but unwarranted precision.

Illustrations of the importance of the ras-oncogene in the B6C3F1 mouse followed. Liver tumors from mice exposed to vinyl chloride have a high frequency of ras-oncogene activation. Those associated with methylene chloride, trichloroethylene, and dichloroacetic acid are similar in ras activation patterns to nonexposed controls. Ras-oncogene activation is not detected in liver tumors associated with the administration of tetrachloroethylene, chloroform, or phenobarbital. Furan produces some novel ras mutations. Maronpot suggests if specific types of oncogene mutation and activation are found in both animal and human tumors that those findings would be important for risk assessment.

Characterization of the cell proliferative response has a number of pitfalls and limitations. Examples were given where cellular proliferation appears to be an important consideration such as in the kidney with respect to d-limonene and unleaded gas and in the bladder for saccharin, but the hepatic-cell proliferation after methylene chloride increased only slightly at the 12-month interval and not at all at the 3-,6-, and 18-month intervals. There is an important temporal relationship between cancer and cellular proliferation but that by itself is not evidence of a causal association.

During the ensuing discussion, S.H. Moolgavkar assured the audience that previous representation of hereditary cancers was consistent with his concept of the two-stage model. Knudson suggested that childhood cancers such as retinoblastoma may represent genetically altered embryonic cells with unconditional cell proliferation. Cancer of the mid-years (osteosarcoma, small cell cancer of the lung, breast cancer) may represent genetic hits—embryonal or not—but with conditional control of cellular proliferation. This control could, in the case of breast cancer for example, be hormonal. Then, it might be that late-age cancers arise from normally dividing tissue in which genetic injury occurs and cell proliferation is enhanced.

TWO-STAGE MODEL
OF CLONAL EXPANSION

The next speaker, S.H. Moolgavkar, described the two-stage model of clonal expansion. He noted that from the data presented by Knudson, it is fairly well established that there are two rate-limiting events for retinoblastoma, loss of two antio-ncogenes. For other tumors, particularly adult cancers, the process described by Knudson is more complex, but Moolgavkar stated those data are consistent with their being two rate-limiting and necessary events on the pathway to malignancy. Further, observations of the appearance of tumors in populations of people or animals, providing cell division kinetics are taken into account, are consistent with two necessary steps.

For risk assessment, we need models that relate exposures to the agents of interest to the concentration of the active metabolite in tissue of interest. Secondly, we need models that relate the microdosimetry (interaction of metabolites with macromolecules, for example) with macrodosimetry (tumor formation). Because risk assessment involves extrapolation outside the range of data, the model needs to be at least approximately correct for accurate extrapolation.

Models may have biological or mathematical misspecifications. In describing the Armitage-Doll model and its limitations, Moolgavkar concluded that this model as currently used ignores the fact that cell division and differentiation are likely to be important in carcinogenesis. Also, the waiting time from stage to stage may differ from exponential distributions (mathematical misspecifications likely). Moreover, the approximations may be useful for epidemiologic studies but do not hold when the probability of tumor is high as in animal studies.

The Moolgavkar two-mutation model postulates two rate-limiting steps called initiation and conversion, represented by irreversible hereditary transitions from normal cells to intermediate cells to cancer cells. In addition to the conversion rates, each cell population has birth and death processes which affect the clonal expansion rates. The ratio of the death rate to the birth rate is the probability that a fraction of initiated cells does not give rise to foci. Moolgavkar postulates that when more than two events are described, as for skin carcinogenesis in mice or for colon tumors in people, only two events may be necessary for the occurrence of the malignant cell and that the other events simply provide a growth advantage (increasing the probability of transformation by increasing the number of target cells for the second event or increasing progression and metastasis after transformation has occurred). The model has been used for human breast and lung cancers and for retinoblastoma.

In presenting examples of applications of the model (radon and lung cancer in rats; N-nitrosomorpholine and liver foci in rats), Moolgavkar emphasized that with this model the shape of the incidence curve is determined by tissue growth and differentiation in contrast with the Armitage-Doll model where the age-specific incidence curve is determined by the number of stages required for malignant transformation. Both examples provided estimates of initiating and converting (promoting) potencies that can be expressed as the proportionate increase per unit dose over background.

The data needs for application of the model include labeling indices for putative intermediate cells at several time points (serial sacrifice studies). Also needed are better models for the cell cycles. The model assumes, for example, that cells divide and die with exponential waiting times and that all cells in the intermediate foci are in the active dividing stage.

The planned formal discussion ensued. J. Wilson noted that Drs. Knudson and Moolgavkar had brought together two competing theories of carcinogenesis—that mutations lead to cancer and cancer is an adaptive response. He suggested that the inability of current assays to identify initiated cells and to approximate the increased cell number with sufficient sensitivity would be a continual problem. R.J. Sielken reminded us that the components of exposure assessments are probability distributions. For risk assessment he advocated considering the variety of estimates that are generated from the use of several forms of two-stage models. T. Thorslund thought that the two-stage model is a desirable start on a new way of estimating risks in the regulatory process. He indicated that the data required by Moolgavkar are rarely available and that considerable training is required to use the model.

APPLICATION OF THE
TWO-STAGE MODEL TO ANIMAL DATA

S.M. Cohen presented the third major address of the workshop based largely on his own research. Cohen initially indicated that he feels that the current bioassay procedure was a good way to screen for carcinogens (he doesn't know a better way), but that it does not provide sufficient information on the mechanism of action to be useful in biological-based cancer risk assessment. He and his colleagues undertook the problem from the engineering simulation approach.

Cohen agreed fundamentally with the two-stage model described by Moolgavkar. The dividing cell has the greatest susceptibility of a genetic mistake and if cancer arises from genetic mistakes the two factors that influence tumorigenesis are increasing the rate at which genetic events occur or increasing the number of times a given cell divides. The initiated and transformed populations of cells can be augmented in size or in proliferative rates by genetic or nongenetic factors.

Noteworthy aspects of the mouse ED01 study of 2-acetylaminofluorene (2-AAF) were outlined. In the liver, the development of cancer is linear, whereas it is not linear in the bladder, but there appears to be a linear dose-response relationship with respect to DNA adduct formation in both the liver and bladder. This occurs because of a difference in the pharmacokinetics related to the development of tumors at the two different sites. In the liver, 2-AAF is metabolized to the N-hydroxyl intermediate and then to the reactive sulfur-containing metabolite. Because the initiated cells or the "cells in the foci" apparently do not metabolize 2-AAF, only the first genetic event occurs in the liver. There is no compound-induced increase in cell proliferation. In contrast, in the bladder the N-hydroxylation occurs as in the liver, but then an N-glucuronide is formed. This glucuronide is excreted into the urine where it is hydrolyzed and can lead to DNA adduct formation in the urothelium. Thus in the bladder, the reactive metabolite causes both initiation and proliferation. Proliferation is only observed above 60 ppm, although DNA adduct formation occurs at dietary concentrations as low as 5ppm. Proliferation in the bladder appears to be essential for tumor formation, however. In conclusion, a two-stage model may be used for either case, but understanding the pharmacodynamics or oncodynamics of the particular tumor-target site is a prerequisite.

Saccharin, which causes bladder cancer in the rat but not in the mouse, hamster, or nonhuman primate, served as an example of a nongenotoxic agent. The male rat is affected to a greater degree than the female. Several types of information were offered to support the importance of proliferation in the development of bladder tumors with saccharin. First, bladder tumors develop principally if the material is given early in life, when cell division in the bladder is normally high. The labeling index (indicator of cell proliferation) is 10 percent at birth, 1.5 percent at seven days, and 0.1 percent at 21 days of age in the rat. Saccharin, co-administered with a cell-proliferating agent (promoter), increases tumor production. Thus, an increase in cell division in the bladder, by irritation as from saccharin or by a promoter, can result in bladder cancer.

The formation of crystals in the urine is the second factor in saccharin induced tumors. The solubility of the saccharin salt is an important aspect in tumor development and can be dependent on the acidity (pKa) of the salt. If the urine is made acid, crystals are not formed in the urine and tumors do not develop. If alkaline, saccharin leads to the formation of silicate crystals in the urine. These crystals irritate the bladder epithelium to cause an increase of cell-turnover rate. This increased proliferative rate is considered responsible for increasing the rate of spontaneous mutation and thus for the induction of bladder cancer.

This sequence of events does not occur with significant frequency in the female rat, nor at all in the mouse. In the mouse, the pH of the urine is not changed. Presumably, in nonhuman primates and humans at doses that are not otherwise toxic, the increase in cellular division would not occur. Several other materials, melamine and uracil, that cause an increase in bladder epithelial cell division by forming crystals (although the crystals are of a different nature) were cited additionally to support the apparent relationship between bladder cancer and urinary crystals.

With other nongenotoxic carcinogens, such as dioxin, it is important to determine whether they mediate cancer through a receptor mechanism. Such a mechanism can induce cancer by increasing cell proliferation as in the case of a hormonal-related receptor, or it may play a role in affecting the immune system via a receptor site on the human-leukocyte antigen (HLA).

In conclusion, just as an indication of mutagenicity does not necessarily indicate that a material is carcinogenic, neither does the ability of an agent to produce cell proliferation at a high dose indicate that it will be carcinogenic (or carcinogenic at lower doses). Therefore, the determination of the dose-relationship to cellular proliferation is important when considering the likelihood of risk.

A regulator's point of view was presented by W. Farland of the USEPA. He said that risk assessment is not merely the estimation of the risk of cancer. The models presented are useful in considering both the quantitative and qualitative assessment of risk and may lead to the establishment of situational and conditional carcinogenic risks. Conditional carcinogens are those that could cause cancer at some dose, whereas situational carcinogens are those that cause cancer only under certain circumstances. Thus, situational carcinogens are important only if the situations under which they cause cancer are likely to occur as a result of conditions. The issues of benign tumors, target specificity, or other mechanisms of action (e.g., decreased immune surveillance) and anti-carcinogenesis may have a place in consideration of the two-stage model. Finally, he indicated that the EPA has considered biologically-based modeling in those few cases where sufficient information is available—particularly in the area of characterization of risk.

C. Barrett followed and commented from the viewpoint of a molecular oncologist on the multiple causes of cancer, suggesting that many mechanisms may be involved. A chemical might cause cancer by inducing a heritable mutation on one critical gene, by inducing heritable epigenetic changes in critical genes, or by clonal expansion of one heritable alteration. Additionally, he suggested that one compound might cause cancer through several pathways.

To emphasize the complexities, Barrett said that if the same processes are operating in humans and rodents but rodent tumors have shorter latent periods, then either there are fewer steps or the rates of transitions from step to step are faster in rodents than in humans. Individual tumors in patients may have anywhere from zero to 10 chromosomes showing loss of heterozygosity (loss of suppressor genes); that is, tumors develop individually.

Three ways by which a substance can influence a multistep carcinogenic process are (1) it can induce heritable mutations in the critical genes (directly or indirectly), (2) it can cause heritable epigenetic changes in critical genes, and (3) it can cause one heritable alteration that increases clonal expansion. Substances acting late in the process may be producing secondary mutational events rather than clonal expansion. Adaptation and potentiation must also be taken into account.

Barrett also cautioned us of the difficulties of generating dose-response curves for mitogenesis and of defining mutagenesis. Cell-cycle control genes and genetic instability are as yet little understood but potentially important, as is also transcriptional control. He concluded that cancer is multicausal, multistep, multigenic, and probably multimechanistic.

In the subsequent general discussion, W. North remarked on the richness of possible modeling approaches and Moolgavkar agreed that we should continue to use the old models until we have more experience with the new ones. K. Crump pointed out that the models would be more helpful if we had comparable data in humans.

Appendix B

Workshop Program

Thursday, November 8, 1990

8:30   **Welcome:** Donald Mattison, University of Pittsburgh
Committee Vice-Chairman

8:35   **Introduction and Objectives:** Richard Griesemer, NIEHS
Workshop Chairman

8:45   **Biological Factors in Two-Stage Models**
Presenter: Alfred Knudson, Fox Chase Cancer Center
9:30   Discussant: Robert Maronpot, NIEHS
9:50   Questions and Comments

10:10   Break

10:30   **Two-Stage Clonal Expansion Model of Carcinogenesis**
Presenter: Suresh Moolgavkar, Fred Hutchinson Cancer Research Center
11:15   Discussants: James Wilson, Monsanto
Robert Sielken, Sielken, Inc.
Todd Thorslund, Clement Associates, Inc.
12:00   Questions and Comments

12:15   Lunch

1:15   **Application of the Two-Stage Model to Animal Data**
        Presenter: Samuel Cohen, University of Nebraska
                      Medical Center
2:00   Discussants: William Farland, EPA
                      Carl Barrett, NIEHS
2:30   Questions and Comments

2:45   **General Discussion**
        Leader: Richard Griesemer, NIEHS
        Issues:
        • The scientific basis for two-stage models of carcino-
          genesis.
        • Can two-stage models of carcinogenesis adequately
          represent biological processes that may involve a
          variety of mechanisms?
        • The implications for the design of rodent bioassays.
        • The implications for health risk assessment.

3:30   Adjourn

Appendix C

Workshop Federal Liaison Group

Murray S. Cohn
Consumer Product Safety
Commission
Bethesda, MD

Joseph Cotruvo
U.S. Environmental Protection
Agency
Washington, DC

William H. Farland
U.S. Environmental Protection
Agency
Washington, DC

Henry S. Gardner
U.S. Army Biomedical Re-
search and Development
Laboratory
Frederick, MD

Herman Gibb
U.S. Environmental Protection
Agency
Washington, DC

Peter Infante
U.S. Department of Labor/
OSHA
Washington, DC

Ronald J. Lorentzen
U.S. Food and Drug Adminis-
tration
Washington, DC

Edward Ohanian
U.S. Environmental Protection
Agency
Washington, DC

Lorenz R. Rhomberg
U.S. Environmenal Protection
Agency
Washington, DC

Robert Scheuplein
U.S. Food and Drug Administration
Washington, DC

Michael Slimak
U.S. Environmental Protection Agency
Washington, DC

Janet A. Springer
U.S. Food and Drug Administration
Washington, DC

Leslie T. Stayner
National Institute for Occupational Safety and Health
Cincinnati, OH

Appendix D

Workshop Attendees

Topic Group Members

Dr. Richard A. Griesemer,
 Chair
National Institute of Environ-
 mental Health Sciences
Research Triangle Park, NC

Dr. Paul T. Bailey
Mobil Oil Corporation
Princeton, NJ

Dr. Kenny S. Crump
Clement Associates, Inc.
Ruston, LA

Dr. Michael A. Gallo
Robert Wood Johnson Medical
 School
University of Medicine and
 Dentistry of New Jersey
Piscataway, NJ

Dr. Donald Mattison
Graduate School of Public
Health
University of Pittsburgh
Pittsburgh, PA

Dr. Franklin E. Mirer
Health and Safety Department
United Auto Workers
Detroit, MI

Dr. D. Warner North
Decision Focus, Inc.
Los Altos, CA

NRC Staff

Dr. Kathleen R. Stratton
Project Director

Dr. Richard D. Thomas
Principal Staff Scientist

Dr. Robert P. Beliles
Senior Staff Scientist

Ms. Linda V. Leonard
Project Assistant

Appendix E

Workshop Organizing Task Group

Richard A. Griesemer
(Chairman)
National Institute of
  Environmental Health
  Sciences
Research Triangle Park, NC

John C. Bailar, III
McGill University School of
  Medicine
Montreal, Canada

Paul T. Bailey
Mobil Oil Corporation
Princeton, NJ

Kenny S. Crump
Clement Associates, Inc.
Ruston, LA

Michael A. Gallo
University of Medicine and
  Dentistry of New Jersey
Piscataway, NJ

Daniel Krewski
Health and Welfare Canada
Ottawa, Ontario
Canada

Donald Mattison
University of Pittsburgh
Pittsburgh, PA

Franklin E. Mirer
United Auto Workers
Detroit, MI

D. Warner North
Decision Focus, Inc.
Mountain View, CA

Issues in Risk Assessment

A Paradigm for
Ecological Risk Assessment

1

Introduction

In 1989, the Committee on Risk Assessment Methodology was convened within the Board on Environmental Studies and Toxicology of the Commission on Life Sciences, National Research Council (NRC) to identify and investigate important scientific issues in risk assessment. The committee was asked to consider changes in the scientific foundation of risk assessment that have occurred since the 1983 report, *Risk Assessment in the Federal Government: Managing the Process*, and to consider applications of risk assessment to noncancer end points.

This report addresses one of the first issues selected by the committee: The development of a conceptual framework for ecological risk assessment, defined as the characterization of the adverse ecological effects of environmental exposures to hazards imposed by human activities. Adverse ecological effects include all environmental changes that society perceives as undesirable. Hazards include unintentional hazards, such as pollution and soil erosion, and deliberate management activities, such as forestry and fishing, that often are hazardous either to a managed resource itself or to other components of the environment. The committee believes that a general framework analogous to the human health risk assessment framework described in the NRC's 1983 report is needed to define the relationship of ecological risk assessment to environmental management and to facilitate the development of uniform technical guidelines. A framework for ecological risk assessment could, for example, be used for the following:

• Evaluation of the consistency and adequacy of individual assessments.
• Comparison of assessments for related environmental problems.
• Explicit identification of the connections between risk assessment and risk management.
• Identification of environmental research topics and data needs common to many ecological risk assessment problems.

Ecological risk assessment is an extraordinarily diverse field whose practitioners include ecologists, fish and wildlife biologists, toxicologists, and pollution-control engineers. Many of the practices in these different fields have grown somewhat independently for many decades, and it was not clear to the committee whether diverse traditions could be united by a common conceptual framework. The committee chose to investigate the feasibility issue by conducting a workshop in which six case studies representing different types of current assessments would be examined with respect to their consistency with a common framework. The six case studies were:

• Assessing the effects of tributyltin on Chesapeake Bay shellfish populations.
• Testing agricultural chemicals for effects on avian species.
• Predicting the fate and effects of polychlorinated biphenyls (PCBs) and 2,3,7,8-tetrachlorodibenzo-*p*-dioxin (TCDD).
• Quantifying the responses of northern spotted owl populations to habitat change.
• Regulating species introductions.
• Harvesting the Georges Bank multispecies fishery.

A workshop on ecological risk assessment was held on February 26-March 1, 1991, at Airlie House, Warrenton, Virginia. The workshop summary (Appendixes C-H) contains summaries of the plenary presentations, case studies and discussions, and breakout sessions. The workshop summary provides much of the supporting information for the conclusions and recommendations presented.

A consensus emerged at the workshop that an ecological version of the 1983 framework is desirable and feasible, but no specific endorsement of a particular framework was sought or obtained. Workshop participants noted several deficiencies in the 1983 framework that prevent direct application to ecological risk assessment. On reviewing the written materials produced at the workshop, the committee concluded that those deficiencies are relevant to health risk assessment as well. The committee chose to respond by modifying the 1983 framework to account for these perceived deficiencies. The committee believes that with modifications, a single framework can accommodate human health and ecological risk assessment.

The committee was not charged with conducting an in-depth analysis of scientific issues in ecological risk assessment or to recommend specific technical guidelines. Many such issues were identified at the workshop, and discussion summaries included in Appendices C-H should provide valuable material for future expert committees charged with evaluating the scientific basis of ecological risk assessment and developing inference guidelines.

Chapter 2 of this report defines the broad uses of ecological risk assessment and its relevance to environmental decision-making at the levels of the individual program, the agency, and society at large. Chapter 3 presents the unified health/ecological risk assessment framework developed by the committee. Chapter 4 highlights key scientific problems limiting the application of ecological risk assessment. The committee's conclusions and recommendations are presented, respectively, in chapters 5 and 6.

For readers interested in further information on topics discussed in this report and its appendices, three of the case studies presented at the workshop were subsequently published in *Environmental Science and Technology* (Fogarty et al.; 1992, Huggett et al., 1992; Kendall, 1992). After the workshop, the U.S. Environmental Protection Agency published a "Framework for Ecological Risk Assessment" (EPA, 1992) that is similar in concept to the framework recommended in this report, although slightly different in terminology and definitions.

2

Scope of Ecological Risk Assessment

All the case studies and most of the discussion at the workshop focused on technically sophisticated assessments performed in narrowly defined regulatory contexts. The scope of ecological risk assessment within the general environmental decision-making process is much broader. The committee recognizes a hierarchy of types of risk assessments, each with its own characteristics. The basic focus of this report is risk assessment in support of day-to-day agency decisionmaking, as exemplified by the case studies. Ecological risk assessments are driven by specific laws or regulations with carefully circumscribed objectives, are science intensive, and provide the principal focus for the quantitative assessment methods and research needs identified in this report.

Risk assessment has a clear role in strategic planning and priority setting. Several plenary session speakers addressed the need for higher-level risk assessments that assist agencies in their planning process or help society to determine its environmental priorities. The U.S. Environmental Protection Agency's (EPA) Relative Risk Reduction Project (EPA, 1990) was cited as an example of such an assessment. The purpose of these assessments is to set priorities and define budgets, and they can be used within agencies and as a means of setting priorities between agencies. Assessments at this level are based principally on expert judgment, rather than on quantitative analysis, but they can benefit from use of an explicit risk assessment framework to organize information and present results in a form useful for decisionmaking.

There is a less obvious, but no less important, role for risk assessment in the process by which society establishes environmental goals. During the closing plenary session, the relationship of ecological risk assessment to goal setting was discussed: Is a particular goal (e.g., preservation in a pristine state) implicitly part of ecological risk assessment, or is risk assessment a value-free tool for transforming politically determined goals into functioning regulations? These issues are important, and the committee believes that it is appropriate and necessary to address them in this report.

The committee emphasizes the need to maintain the clear conceptual separation between risk management and risk assessment. Goal setting is a risk-management function; therefore, the definition of ecological risk assessment cannot contain implicit ecological preservation or restoration goals. The committee agrees, however, that risk assessment can play a vital communication function in goal setting. During the workshop, Dr. Lovejoy, from the Smithsonian Institution, noted that society must define its goals in a scientifically informed way and suggested that ecological risk assessment should play an educational role in this process. Dr. Yosie, from the American Petroleum Institute, touched on the issue of goal setting and suggested a role for risk assessment in clarifying public debates over environmental policy by making explicit the environmental consequences of particular policy choices. This process is continuous in the United States and worldwide, as exemplified by the climate-change debate and by the current discussion of the idea of sustainable development. Risk assessment clearly can help by providing operational definitions of generally understood but vaguely defined concepts, such as sustainability, by identifying the scientific information needed to evaluate policy alternatives, and by delineating the consequences of particular choices.

The definitions, research areas, conclusions, and recommendations discussed in the remainder of this report are intended to lay the foundation for an approach to ecological risk assessment that can contribute to environmental decisionmaking at all levels, for all types of environmental problems.

3

Revision of 1983 Framework To Incorporate Ecological Risk Assessment

COMPONENTS OF THE 1983 FRAMEWORK

Risk Assessment in the Federal Government: Managing the Process (NRC, 1983), often called the "Red Book," proposed a conceptual framework for risk assessment that incorporates research, risk assessment, and risk management (Figure 3-1). Risk assessment was defined as "the characterization of the potential adverse health effects of human exposures to environmental hazards." The overall scheme and terminology proposed in the 1983 report entailed hazard identification, dose-response assessment, exposure assessment, and risk characterization. Hazard identification was defined as "the process of determining whether exposure to an agent can cause an increase in the incidence of a health condition," including "characterizing the nature and strength of the evidence of causation." Dose-response assessment was defined as "the process of characterizing the relation between the dose of an agent administered or received and the incidence of an adverse health effect . . . as a function of human exposure to the agent," accounting for exposure intensity, age, sex, lifestyle, and other variables affecting human health responses to hazardous agents. Exposure assessment was defined as "the process of measuring or estimating the intensity, frequency, and duration of human exposures to an agent currently present in the environment or of estimating hypothetical exposures that might arise from the release of new chemicals into the environment." Risk characterization was defined

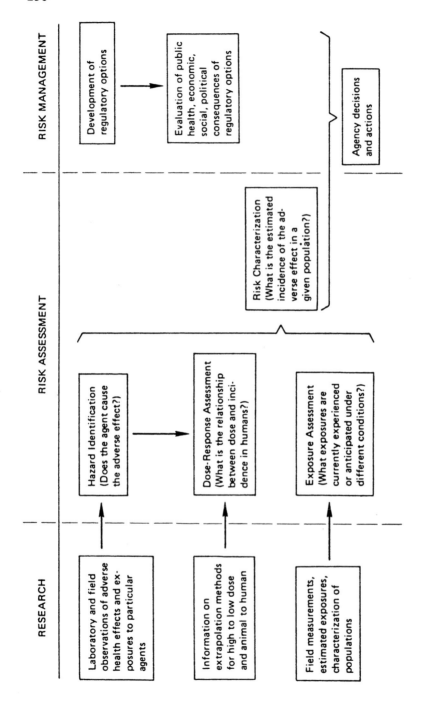

FIGURE 3-1   Elements of risk assessment and risk management.

as "the process of estimating the incidence of a health effect under the various conditions of human exposure described in exposure assessment. It is performed by combining the exposure and dose-response assessments. The summary of effects of the uncertainties in the preceding steps are described in this step."

The 1983 report further defined points in the risk assessment process ("components") where inferences must be made and scientifically plausible options ("inference options") from which a risk assessor must choose regarding those components. The report did not, however, include in-depth discussion of scientific issues in health risk assessment. The 1983 committee's objectives were limited to addressing institutional and procedural issues: whether the analytic process of risk assessment should be cleanly separated from the regulatory process of risk management, whether a single organization could be designated to perform risk assessments for all regulatory agencies, and whether uniform risk assessment guidelines could be developed for use by all regulatory agencies. The general framework for health risk assessment developed by the 1983 committee (Figure 3-1) was intended to define the boundaries between risk assessment and risk management and to facilitate the development of uniform technical guidelines. The committee recommended that a board on risk assessment methods be established and assigned the tasks of assessing the scientific basis of risk assessment, establishing inference guidelines, evaluating agency experiences with risk assessment, and identifying research needs in risk assessment.

CONSISTENCY OF CASE STUDIES
WITH THE 1983 FRAMEWORK

Most of the case studies fit reasonably well into the 1983 framework, although the relative emphasis on the four components of risk assessment varied considerably among the studies. The three case studies dealing with environmental chemicals provided the most obvious fits. All three included discussions of hazard identification, defined as determination of the physical, chemical, and toxic effects of the substances or stresses being examined. They differed substantially in their balance between data and models, but a fairly clean distinction could be drawn between exposure assessment (patterns of contamination in time and space, exposure, and doses) and dose-response assessment (quantitative relation of exposures to toxic effects).

The Georges Bank study appeared to be the most complete of the six. The determination of the qualitative effects of fishing on population and community dynamics is clearly analogous to the determination of contaminant effects and can legitimately be called "hazard identification." Estimates of fishing effort and models of the responses of populations to exploitation are equivalent to exposure and dose-response assessment of chemicals. The expression of outcomes in terms of likely future population sizes and yields carries risk characterization several steps further than was done in any of the contaminant studies.

The spotted owl study focused on only one aspect of the assessment process: estimation of basic demographic characteristics of spotted owl populations. However, other published work on the spotted owl (Dawson et al., 1986; Salwasser, 1986) available to the committee relates forest-cutting patterns to population dynamics and clearly includes exposure and dose-response assessments in the sense in which these terms are used in the Red Book framework.

The species-introduction case study does not appear at first to fit the standard definition of a risk assessment. No scientific principles or decision criteria were presented at the workshop, although theoretical work was described in some of the breakout sessions. The consensus among participants in the workshop was that the procedure used by the U.S. Department of Agriculture (USDA) to evaluate proposed species introductions is not risk assessment. The committee believes, however, that USDA's process fits within the general definition of hazard identification as presented in the 1983 report. The objective appears to be to collect enough information to determine whether a proposed introduction constitutes a hazard to the environment. If no hazard is found, the introduction proceeds. The USDA process might more accurately be described as safety assurance.

One weakness in all the case studies was inadequate risk characterization. Only one of the case studies, the Georges Bank study, included any quantification of risks in terms that could be used for risk-benefit calculations, valuation studies, or other quantitative comparisons applicable to decision-making. Even in this case, the value of the assessment to decision-making is uncertain. During the plenary discussion, the author of the study emphasized that communication between scientists and managers is still inadequate and that fisheries management actions are often only marginally influenced by quantitative assessments.

The committee notes that risk characterization is the least-developed component of the 1983 framework. In the 1983 report, risk characterization is defined simply as an integration of exposure and dose-response information. It seems clear from the 1991 workshop that effective ecological risk characterization is more than an exercise in arithmetic. Many of the results presented at the workshop have no immediate relevance to decision-making and mean little or nothing to the public. The procedure used in pesticide registration, as described in the agricultural-chemical case study, provides an excellent example. The method used is to compare doses that cause death or impairment of standard test birds with estimated exposures in typical applications. On the basis of the comparison, the risk manager is expected to make a decision about the environmental acceptability of the pesticide being considered. No attempt is made to account for interspecies differences, to assess the threat to the viability of wild avian populations, to estimate the fraction of the landscape that might be affected, or to quantify the value of the wildlife that might be lost.

Ecological risk assessments have no equivalent of the lifetime cancer-risk estimate used in health risk assessment. The ecological risks of interest differ qualitatively between different stresses, ecosystem types, and locations. The value of avoiding these risks is not nearly as obvious to the general public as is the value of avoiding exposure to carcinogens. Because few risk managers are trained as ecologists, effective communication between risk managers and technical staff is essential in sound risk-management decisions.

Approaches to hazard identification exemplified in the case studies were, on the other hand, substantially more diverse and in some cases more sophisticated than envisioned in the 1983 framework. The 1983 definition of this component was limited to scientific inferences about whether specific effects, such as cancer, were causally associated with specific chemical substances. Identification of ecological hazards also includes identification of specific species or ecosystems of interest, delineation of study areas, and determination of types of laboratory or field data on which an assessment will be based. These decisions reflect both scientific considerations (which systems are vulnerable? what kinds of effects are possible?) and management considerations (which species

or ecosystems are to be protected? must costs be weighed against benefits? is the objective to protect the resource or to optimize exploitation
of the resource?). The committee agrees with the consensus from the
workshop that the initial phases of an ecological risk assessment involve
a consideration of regulatory/legal mandates that goes well beyond the
definition of hazard identification presented in the 1983 report.

INTEGRATION OF ECOLOGICAL RISK
INTO THE 1983 FRAMEWORK

The committee believes that integration of ecological risks into the
1983 risk assessment framework is preferable to developing a de novo
ecological risk assessment framework. Like health risk assessment,
ecological risk assessment must be defined in broad terms if it is to be
applicable to the full array of environmental problems that regulatory
and resource management agencies must address. Moreover, any framework chosen for ecological risk assessment must be simple, flexible, and
general, so that it will be understood by both scientists and the risk
managers with whom scientists must communicate. The 1983 framework, by any measure, has been extraordinarily successful in communicating the broad features of health risk assessment throughout the scientific and regulatory communities. Although ecological risk assessment
and human health risk assessment differ substantially in terms of scientific disciplines and technical problems, the committee believes that the
underlying decision process is the same for both. The function of risk
assessment is to link science to decision-making, and that basic function
is essentially the same whether risks to humans or risks to the environment are being considered. Finally, the committee believes that prospects for integration of human and ecological concerns into comprehensive environmental policies protective of both will be enhanced if a
common framework and terminology can be found that describes both
kinds of risk assessments.

The committee agrees with the consensus at the workshop that the
framework defined in the 1983 report is inadequate as written for application to ecological problems because the framework (1) does not account for legal mandates and other policy considerations that substantially influence the initial stages and focus of ecological risk assessments

and (2) pays insufficient attention to the critical problem of effective communication with risk managers and the public. The opinion of the committee, however, is that these deficiencies are not unique to ecological risk assessment. Differences in the functions of different regulatory agencies clearly influence the types of data and inference guidelines used in health risk assessments, and effective risk communication is as important (and often as inadequately performed) in health as in ecological risk assessment.

DEFINITION OF FRAMEWORK COMPONENTS FOR ECOLOGICAL RISK ASSESSMENT

Hazard identification is redefined to be the determination of whether a particular hazardous agent is associated with health or ecological effects that are of sufficient importance to warrant further scientific study or immediate management action.

This change in definition is intended to account for the influence of regulatory mandates and other policy considerations on the conduct of risk assessments. Examples of such influences are restrictions on data acquisition or response time (e.g., premanufacture notification assessments under the Toxic Substances Control Act), standardized data requirements and regulatory criteria (pesticide registration under the Federal Insecticide, Fungicide, and Rodenticide Act), and the scoping provisions of the National Environmental Policy Act. Other aspects of hazard identification, such as investigation of cause-effect relationships and preliminary screening, would remain essentially unchanged.

Exposure-response assessment is defined as the determination of the relation between the magnitude of exposure and the probability of occurrence of the effects in question. Replacement of the term "dose" with a more general term is required, because "dose" has a distinctly medical connotation and cannot be effectively applied to nonchemical stresses, such as habitat change or harvesting. The "responses" addressed in ecological risk assessments include direct effects of exposure and the much broader indirect effects, such as secondary poisoning of raptors due to accumulation of pesticide residues in their prey and effects of harvesting on fish-community structure.

Exposure assessment is defined as the determination of the extent of exposure to the hazardous agent in question before or after application of regulatory controls. In the committee's view, the term "exposure" can legitimately be applied to nonchemical stresses, including physical stresses (such as habitat and UV radiation) and biological stresses (such as species introductions). The committee considered changes in terminology on the grounds that the term "exposure" is too closely associated with chemical risks. However, the alternative terms discussed (e.g., stress and stressor) were unsuitable because of conflicts with medical uses of the same or similar terms.

Risk characterization is defined as the description of the nature and often the magnitude of risk, including attendant uncertainty, expressed in terms that are comprehensible to decision-makers and the public. Extension of the definition provided in the 1983 report is needed to permit more explicit discussion of uncertainty, to facilitate expression of risks in management-relevant terms (including valuation), and to emphasize the importance of communication between scientists and managers. The committee believes that improved communication is as important for health risk assessment as it is for ecological risk assessment.

The revised framework is summarized in Figure 3-2. The relationships among the four components are unchanged from the Red Book: hazard identification is the initial step in an assessment. Exposure assessment and exposure-response assessment occur roughly in parallel and must be closely linked. The arrangement of those components in Figure 3-2, within a single box divided in half by a "permeable membrane," is intended to emphasize the ties between them. Risk characterization synthesizes the results of technical analyses and expresses them in a form suitable for valuation studies or other policy analyses that are carried out as part of risk management.

In addition to the four basic components, Figure 3-2 depicts two aspects of risk assessment that the committee wants to emphasize. As previously noted, it is essential to recognize that management considerations (e.g., regulatory constraints on the scope or time available for an assessment or legally prescribed definitions of acceptable or unacceptable uses) can shape the hazard-identification step. The committee would also like to emphasize the need to create a connection between the results of today's risk assessments and the science base for future risk assessments. The risk assessment process should not end when a regulatory decision is made. Followup in the form of monitoring (where

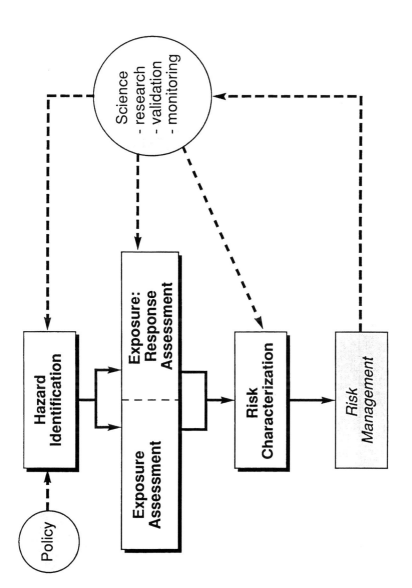

FIGURE 3-2   Extention of the 1983 NAS risk assessment paradigm to include ecological as well as human health risks.

measurable effects have been predicted), validation studies, and basic research are needed to improve the data and models available to technical risk assessors whenever the same or a similar problem is encountered in the future.

4

Key Scientific Problems Limiting Application of Ecological Risk Assessment

EXTRAPOLATION ACROSS SCALES

The most common scientific limitation exemplified in the case studies is the problem of extrapolating across scales of space, time, and ecological organization. For the most part, scientific data related to a specific stressor are limited to what can be obtained in a controlled laboratory setting or in a limited field study. Observations of environmental contamination and ecological effects of tributyltin were limited to a few marinas. Testing of pesticides even in the best of circumstances is limited to small field plots and carefully controlled applications. Table 4-1 shows, for all the case studies, the scales at which the data used in the assessments were collected and the scales of interest in decision-making. In most cases, the scales of interest in decision-making are substantially larger in space and of longer duration than could be accommodated in any practical assessment effort. Some form of extrapolation, either with explicit mathematical models or with judgment-based decision rules, is necessary to make the risk assessments useful for decision-making. The PCB study discussed by Di Toro (Appendix E) clearly illustrated the value of explicit models for estimating recovery times in response to hypothetical management actions. In the pesticide registration process described by Kendall (Appendix E), extrapolation is based primarily on qualitative evaluation of test data and information on expected use patterns. Kendall argued that models of ecological effects of pesticides are needed to reduce uncertainty and to account for effects

TABLE 4-1 Scales of Observation and Management in Case Studies Evaluated by the Committee.

Case Study	Observational Scale		Management Scale	
	Spatial	Temporal	Spatial	Temporal
Tributyltin	$< 1 m^3$ ~ 1 ha	< 1 yr (laboratory) < 5 yr (field)	Chesapeake Bay	$>$ 5 yr
Agricultural chemicals	~ 1 ha	< 1 yr (field test)	Agricultural region	$>$ 5 yr
PCB and TCDD	< 1 L	~ 1 month (laboratory)	Lakes or rivers	$>$ 10 yr
Spotted owl	~ 300 km^2	< 6 yr	Pacific northwest	$>$ 100 yr
Species introduction	$< 100 m^2$	~ 1 yr (greenhouse)	Agricultural region	$> >$ 1 yr
Georges Bank	~ $10^4 km^2$	last 30 yr	~ $10^4 km^2$	next 5 yr

that cannot be directly measured in test systems. Extrapolation was not explicitly considered in the paper by Anderson (Appendix E). Other published literature attempts to relate spotted owl abundance to regional patterns of old-growth forest harvesting (Salwasser et al., 1986; Lande, 1988). Bartell et al. (1992) have recently discussed the problem of estimating population and ecosystem-level effects of toxic contaminants from laboratory toxicity tests. A variety of approaches are now being developed for extrapolating local-scale disturbances (e.g., fires or insect outbreaks) to regional-scale changes in landscape patterns (Costanza et al., 1990; Turner and Gardner, 1991; Graham et al., 1991). Extrapolations from spatial and temporal scales suitable for rigorous experiments and observations to scales relevant to environmental management appear to be essential for adequate characterization of ecological risks.

QUANTIFICATION OF UNCERTAINTY

Formal analysis of uncertainty is another major subject for improvement in ecological risk assessments of all types. The "uncertainties" discussion group at the workshop identified three general categories of uncertainty that affect all types of risk assessments:

• *Measurement uncertainties*, e.g., low statistical power due to insufficient observations, difficulties in making physical measurements, inappropriateness of measurements, and natural variability in organic responses to stress.
• *Conditions of observation*, e.g., spatiotemporal variability in climate and ecosystem structure, differences between natural and laboratory conditions, and differences between tested or observed species and species of interest for risk assessment.
• *Inadequacies of models*, e.g., lack of or knowledge concerning underlying mechanisms, failure to consider multiple stresses and responses, extrapolation beyond the range of observations, and instability of parameter estimates.

Measurement uncertainties can be reduced by making more and better measurements. Uncertainties related to conditions of observation cannot be reduced, but often they can be quantified using empirical regression techniques (Suter et al., 1983), time series analysis (Jassby and Powell, 1990), or formal model uncertainty analysis (Bartell et al., 1992). Di Toro and Fogarty et al. provided examples of model uncertainty analyses in their case study papers (Appendix E). Uncertainties related to inadequacies of models (or scientific ignorance in general) are much more difficult to quantify.

Choices between risk assessment methodologies often involve tradeoffs between different types of uncertainty. For example, decisions about the need for pesticide testing are now based on qualitative evaluation of toxicity and exposure data (Urban and Cook, 1986). Explicit models of the effects of toxicant exposure on the abundance and persistence of bird populations have been developed (Grier, 1980; Tipton et al., 1980; Samuels and Ladino, 1983) and could be used to quantify uncertainties related to variability in exposures or extrapolation from field plots to natural landscapes. Relying on expert judgment avoids the need to postulate particular mechanisms of exposure or complex population dynamics but prevents risk assessors from providing information on the value of collecting additional information to reduce uncertainties or providing information on the ecological costs and benefits of regulatory decisions. Using a model to quantify uncertainties would in principle permit more useful risk assessments, but if the model itself is a poor representation of reality, the results might be totally meaningless.

The committee believes that improvements are needed in techniques for qualitative and quantitative analysis of uncertainty for ecological risk assessment. Techniques for model uncertainty analyses developed by systems engineers have been used by ecologists for more than a decade (Gardner et al., 1981; Bartell et al., 1992; Di Toro, Appendix E). The large and growing technical literature on decision analysis (Raiffa, 1970; Lindley, 1985; Von Winterfeldt and Edwards, 1986) has been much less thoroughly exploited (see Walters (1986) and Reckhow (1990) for examples of ecological applications of Bayesian decision theory) and should be surveyed for potentially useful approaches.

VALIDATION OF PREDICTIVE TOOLS

Improvements in the mathematical models, qualitative and quantitative decision rules, and other predictive tools used in ecological risk assessment still are needed. Although the committee refers to the process of improvement as "validation," we recognize that none of the approaches in question can be proved fully valid in the sense of perfectly predicting natural ecosystem behavior under all circumstances. The purpose of validation is to improve the credibility and reliability of predictive methods. Validation must be viewed as an iterative process in which predictions are tested, models are refined, and then new predictions are tested.

At least three kinds of studies can contribute to validation: improved measurements of specific quantities and tests of assumptions, experimental testing of models under reasonably realistic conditions (e.g., ponds or enclosures), and monitoring of ecological effects of dams, power plants, or other projects to determine the accordance between actual effects and effects predicted before construction or operation of the dams. Each kind of study has its own advantages and disadvantages, and all three should be included in validation programs.

This committee is not the first to note the need for validation studies. Similar recommendations can also be found in at least two previous NRC reports (NRC, 1981; NRC, 1986). In spite of virtually unanimous support within the scientific community for this activity, the resources currently being expended for improvement of predictive tools are much smaller than those devoted to repetitive assessments and routine monitoring of compliance with permit requirements. The importance of enhanced validation programs needs to be recognized by all regulatory and resource management agencies.

VALUATION

Valuation and cost-benefit analyses are recognized as integral components of the risk-management process. Such analyses contribute to the regulations that provide the context for risk assessments and to the eventual risk-management decisions. Cost-benefit analyses are major parts of the planning and ranking process within and between agencies. Ecological cost-and-benefit analyses have gained acceptance where individual behavior can be used to directly reflect economic preferences, e.g., recreational use and associated travel-cost analysis (Yang et al., 1984; DesVouges and Skahen, 1985). These analyses have also been applied with some success where people's direct use of a resource was the specific issue, e.g., dam construction vs. maintenance of the natural river channel. Methods also have been developed for monetizing ecological values beyond those associated with the use of a resource, but the uncertainties associated with applications of those methods are often quite high.

Ecological values are sometimes described by resource economists as services provided by the environment to humans. Such economists categorize economic values into two segments termed "use" and "non-use" values. As noted above, reasonably reliable techniques are available for determining use values (e.g., land valuation and recreational use) from the actual behavior of resource users. Several methodologies have been developed to date for estimating non-use values. For example, contingent valuation uses public surveys to elicit statements of how much an individual hypothetically would be willing to pay for improvements (or to prevent reductions) in the quantity or quality of natural resources. It requires people to assign subjectively economic values for environmental goods. Recent empirical research indicates that the results vary depending on the way the assessments are elicited (Opaluch and Segerson, 1989; Grigalunas and Opaluch, 1991; Hausman, 1991; Rosenthal and Nelson, 1992), and the resulting values must be interpreted with care.

Clearly, a considerable need remains for increased communication and clarification between ecologists and economists to improve the use of valuation methods in ecologic risk-management decisions. There is already a substantial literature on the economic value of wetland ecosystems (Scodari, 1990). Valuation of other kinds of ecosystems is being actively discussed (e.g., Orians, 1990), but generally accepted principles for ecosystem valuation do not yet exist.

5

Conclusions

- Risk assessment is defined as a general process for linking science to decision-making. Definitions and concepts for ecological risk assessment can be defined analogously with those provided for human health risk assessment in the NRC's 1983 report, *Risk Assessment in the Federal Government: Managing the Process*. The scope of ecological risk assessment, defined in this way, is not limited to technical analyses supporting day-to-day regulatory decisions. Ecological risk assessment can also contribute to strategic planning within and between federal agencies and to clarifying the ongoing public debate concerning national and international environmental priorities.

- The four-part framework for health risk assessment (hazard identification, dose-response assessment, exposure assessment, risk characterization) described in the 1983 report insufficiently characterizes the interaction between science and management in risk assessment. Management considerations (e.g., regulatory constraints on the scope or time available for an assessment for legally prescribed definitions of acceptable or unacceptable risks) shape the hazard identification step. Communication of risks in terms relevant to decision-making is a critical aspect of risk characterization that was largely unexplored in the 1983 report. (This topic was explored in a subsequent NRC report, *Improving Risk Communication* (NRC, 1989)). Although the deficiencies were initially identified by the committee with respect to ecological risk assessment, the committee believes they are relevant to human health risk assessment as well.

• If the definitions used in the 1983 report are expanded then, with minor changes in terminology, a single framework can accommodate both human health and ecological risks. Hazard identification should be expanded to include identification of policy considerations or regulatory mandates that influence the scope and objectives of an assessment. Risk characterization should be expanded to provide explicit discussion of uncertainty, facilitate expression of risks and management-relevant terms, and emphasize the importance of communication between scientists and managers. Prospects for integration of human and ecological concerns into comprehensive environmental policies that are protective of both will be enhanced if the assessments employ a common framework and terminology.

• Several scientific problems are common to all types of ecological risk assessments. Most of them are now subjects of active scientific research: extrapolation across scales of time, space, and ecological organization; quantification of uncertainty; validation of predictive tools; and valuation, especially quantification, of non-use values.

• Technical guidance is needed on the scientific content of ecological risk assessments. In-depth analysis of the scientific issues involved in specific applications of ecological risk assessment is beyond the charge of this committee. Additional expert committees drawn from the academic, public, and private sectors are needed to address these issues. The workshop summary (Appendix H) contains many specific examples of topics for which guidance is needed; the NRC's report *Ecological Knowledge and Environmental Problem-Solving* (NRC, 1986) provides a useful model for synthesis and presentation of the science relevant to improving ecological risk assessments.

6

Recommendations

• The committee recommends that risk assessors, risk managers, and regulatory agencies should adopt a uniform framework for ecological risk assessment. The framework used should be general and flexible and should facilitate communication between scientists and risk managers. The objective is to systematize the regulatory process, not to specify a particular calculational procedure or data requirement. The extension of the 1983 NRC human health risk assessment framework described in this report and summarized in Figure 3-2 emphasizes the common elements of health risk and ecological risk assessment.

• The committee recommends that state and federal agencies expand the issue of risk assessment in strategic planning and priority setting. Those assessments can help agencies to focus their resources on critical environmental problems and uncertainties, and that focus would result in more cost effective and efficient regulatory initiatives. Assessments at this level can be principally judgment-based and need not involve explicit quantification.

• The committee recommends that agencies support the development of improved methods of risk characterization and consistent guidelines for applying them. Improvement is needed in extrapolation of population and ecosystem effects, expression of risks in terms that are useful for decisionmaking and understood by the public at large, and evaluation and communication of quantitative and qualitative uncertainties.

• EPA and other agencies, including state agencies, should institute systematic followup of risk assessments with research and monitoring to determine the accuracy of predictions and resolve remaining uncertainties.

• The committee recommends that EPA and other agencies support systematic research programs to improve the credibility and utility of ecological risk assessments and draw on scientific expertise available outside the agencies themselves to develop technical guidance on the scientific content of ecological risk assessments.

References

Bartell, S.M., R.H. Garner, and R.V. O'Neill. 1992. Ecological Risk Estimation. Boca Raton, Fla.: Lewis Publishers.

Costanza, R., F.H. Sklar, and M.L. White. 1990. Modeling coastal landscape dynamics. BioScience 40:91-107.

Dawson, W.R., J.D. Ligon, J.R. Murphy, J.P. Myers, D. Simberloff, and J. Verner. 1986. Report of the Scientific Advisory Panel on the Spotted Owl. The Condor 89(1):205—229. (Audubon Conservation Report 7). National Audubon Society, New York, N.Y.

DesVouges, W., and J. Skahen. 1985. Techniquest to Measure Damages to Natural Resources.

EPA (U.S. Environmental Protection Agency, Science Advisory Board). 1990. Reducing Risk. EPA SAB-EC-90-021. U.S. Environmental Protection Agency, Washington, D.C.

EPA (U.S. Environmental Protection Agency). 1992. Framework for Ecological Risk Assessment. EPA 630/R-92-001. U.S. Environmental Protection Agency, Washington, D.C.

Fogarty, M.J., A.A. Rosenberg, and M.P. Sissenwine. 1992. Fisheries risk assessment: Sources of uncertainty. Environ. Sci. Technol. 26:440-447.

Gardner, R.H., R.V. O'Neill, J.B. Mankin, and H.H. Carney. 1981. A comparison of sensitivity analysis and error analysis based on a stream ecosystem model. Ecological Modelling 12:173-190.

Graham, R.L., C.T. Hunsaker, R.V. O'Neill, and B.L. Jackson. 1991.

Ecological risk assessment at the regional scale. Ecol. App. 1:196-206.

Grier, J. 1980. Modeling approaches to bald eagle population dynamics. Wildl. Soc. Bull. 8:316-322.

Grigalunas, T.A., and J.J. Opaluch. 1991. Non-use values under ambivalence: Theory and measurements of ill-defined preference orderings. Paper presented at the Annual Meeting of the American Economics Association, New Orleans, La. January.

Hausman, J.A. 1991. Comments on economic methodologies for damage assessments. Paper presented at the National Oceanic and Atmospheric Administration Workshop on Economic Methodologies for Damage Assessments, Washington, D.C. November 20.

Huggett, R.J., M.A. Unger, P.F. Seligman, and A.O. Valkirs. 1992. The marine biocide tributyltin. Environ. Sci. Technol. 26:232-237.

Jassby, A.D., and T.M. Powell. 1990. Detecting changes in ecological time series. Ecology 71:2044-2052.

Kendall, R.J. 1992. Farming with agrochemicals: The response of wildlife. Environ. Sci. Technol. 26:239-245.

Lande, R. 1988. Demographic models of the northern spotted owl *(Stix occidentalis caurina)*. Oecologia (Berlin) 75:601—607.

Lindley, D.V. 1985. Making Decisions, 2nd ed. New York: Wiley.

Mantel, N., and W.R. Bryan. 1961. Safety testing of carcinogenic agents. J. Nat. Canc. Inst. 27:455—470.

NRC (National Research Council). 1981. Testing for Effects of Chemicals on Ecosystems. Washington, D.C.: National Academy Press.

NRC (National Research Council). 1983. Risk Assessment in the Federal Government: Managing the Process. Washington, D.C.: National Academy Press. 191 pp.

NRC (National Research Council). 1986. Ecological Knowledge and Environmental Problem-Solving: Concepts and Case Studies. Washington, D.C.: National Academy Press. 388 pp.

NRC (National Research Council). 1989. Improving Risk Communication: Washington, D.C.: National Academy Press. 332 pp.

Opaluch, J.J., and K. Segerson. 1989. Rational roots of irrational behavior: New models of economic decision making. Northeast J. of Agric. Resource Eco. 18:2

Orians, G.H., ed. 1990. The Preservation and Valuation of Biological Resources. Seattle, Wash.: The University of Washington Press.

Raiffa, H. 1970. Decision Analysis: Introductory Lectures on Choices under Uncertainty. Reading, Mass.: Addison-Wesley.

Reckhow, K.H. 1990. Bayesian inference in non-replicated ecological studies. Ecology 71:2053-2059.

Rosenthal, D.H., and R.H. Nelson. 1992. Why existence value should not be used in cost-benefit analysis. J. Policy Anal. Mgmt. 11: 116—122.

Salwasser, H. 1986. Conserving a regional spotted owl population. Pp. 227-247 in Ecological Knowledge and Environmental Problem Solving. Washington, D.C.: National Academy Press.

Samuels, W.B., and A. Ladino. 1983. Calculation of seabird population recovery from potential oilspills in the mid-Atlantic region of the United States. Ecol. Model. 21:63-84.

Scodari, P.F. 1990. Wetlands Protection: The Role of Economics. Environmental Law Institute, Washington, D.C.

Suter, G.W., II, D.S. Vaughn, and R.H. Garner. 1983. Risk assessment by analysis of extrapolation error. Environ. Toxicol. Chem. 2:369-378.

Tipton, A.R., R.J. Kendall, J.F. Coyle, and P.F. Scanlon. 1980. A model of the impact of methyl parathion spraying on a quail population. Bull. Environ. Contam. Toxicol. 25:586-593.

Turner, M.G., and R.H. Gardner, eds. 1991. Quantitative Methods in Landscape Ecology. New York: Springer-Verlag.

Urban, J.D. and N.J. Cook. 1986. Ecological Risk Assessment: Standard Evaluation Procedures. EPA 540/9-85-001. U.S. Environmental Protection Agency, Hazard Evaluation Division, Washington, D.C. Available as NTIS PD 86-247-657.

Von Winterfeldt, D., and W. Edwards. 1986. Decision Analysis and Behavioral Research. Cambridge: Cambridge University Press.

Walters, C.J. 1986. Adaptive Management of Renewable Resources. New York: MacMillan.

Yang, E., R. Dower, and M. Menefree. 1984. The Use of Economic Analysis for Valuing Natural Resource Damage.

Appendix A

Workshop Participants

Mr. James Akerman
U.S. Environmental Protection
 Agency
Washington, DC

Dr. David Anderson
U.S. Department of the
 Interior
Fort Collins, CO

Dr. Paul Anderson
ENSR Consulting and Engi-
 neering
Acton, MA

Dr. John C. Bailar, III
McGill University School of
 Medicine
Montreal, Canada

Dr. Paul T. Bailey
Mobil Oil Corporation
Princeton, NJ

Dr. Lawrence Barnthouse
Oak Ridge National Laboratory
Oak Ridge, TN

Dr. Steven Bartell
Oak Ridge National Laboratory
Oak Ridge, TN

Dr. Robert Beliles
National Research Council
Washington, DC

Dr. David Bergsten
U.S. Department of Agriculture
Hyattsville, MD

Dr. Gregory R. Biddinger
Exxon Biomedical Sciences, Inc.
East Millstone, NJ

Dr. Erich W. Bretthauer
U.S. Environmental Protection
 Agency
Washington, DC

Dr. Garry D. Brewer
Yale University
New Haven, CT

Dr. Joanna Burger
Rutgers University
Piscataway, NJ

Dr. Lawrence A. Burns
U.S. Environmental Protection
   Agency
Athens, GA

Dr. James T. Carlton
Maritime Studies Program
Williams College
Mystic, CT

Dr. Raymond Carruthers
U.S. Department of
   Agriculture
Ithaca, NY

Professor Robert Colwell
University of Connecticut
Storrs, CT

Dr. William Cooper
Michigan State University
East Lansing, MI

Dr. Robert Costanza
Chesapeake Biological
   Laboratory
Solomons, MD

Dr. Joseph Cotruvo
U.S. Environmental Protection
   Agency
Washington, DC

Dr. Jack R. Coulson
USDA-ARS-PSI
Beltsville, MD

Dr. William Desvousges
Research Triangle Institute
Research Triangle Park, NC

Dr. Dominic M. Di Toro
Manhattan College
Bronx, NY

Dr. William H. Farland
U.S. Environmental Protection
   Agency
Washington, DC

Mr. John Fedkiw
U.S. Department of Agriculture
Washington, DC

Dr. James Ferguson
National Library of Medicine
Bethesda, MD

Dr. Michael P. Firestone
U.S. Environmental Protection
   Agency
Washington, DC

Dr. James Gagne
American Cyanamid Company
Princeton, NJ

Dr. Hector Galbraith
ENSR Consulting and
   Engineering
Acton, MA

Dr. John H. Gentile
U.S. Environmental Protection
  Agency
Narragansett, RI

Dr. Bernard D. Goldstein
Robert Wood Johnson Medical
  School
Piscataway, NJ

Dr. Cliff Habig
Jellinek, Schwartz, Connolly,
  and Freshman
Washington, DC

Dr. S.G. Hildebrand
Oak Ridge National Laboratory
Oak Ridge, TN

Dr. John E. Hobbie
Marine Biological Laboratory
Woods Hole, MA

Dr. John W. Huckabee
Electric Power Research
  Institute
Palo Alto, CA

Dr. Robert J. Huggett
Virginia Institute of Marine
  Science
Gloucester Point, VA

Mr. Henry Jacoby
U.S. Environmental Protection
  Agency
Washington, DC

Dr. F. Reed Johnson
U.S. Naval Academy
Annapolis, MD

Dr. Jason Johnston
Karch & Associates, Inc.
Washington, DC

Dr. Ronald Kendall
Clemson University
Clemson, SC

Dr. Mary Kentula
U.S. Environmental Protection
  Agency
Corvallis, OR

Dr. Richard Kimerle
Monsanto Company
St. Louis, MO

Dr. Ronald Landy
U.S. Environmental Protection
  Agency
Washington, DC

Dr. Brian P. Leaderer
Yale University School of
  Medicine
New Haven, CT

Ms. Linda Leonard
National Research Council
Washington, DC

Dr. Josh Lipton
RCG/Hagler Bailly Inc.
Boulder, CO

Dr. Gerald Llewellyn
Virginia State Department of
   Health
Richmond, VA

Dr. Orie L. Loucks
Miami University
Oxford, OH

Dr. Thomas E. Lovejoy
Smithsonian Institution
Washington, DC

Dr. J. Larry Ludke
Fish and Wildlife Service
Kearneysville, WV

Dr. Alan W. Maki
Exxon Corporation
Anchorage, AK

Dr. Pamela A. Matthes
U.S. Department of the
   Interior
Washington, DC

Dr. David Mauriello
U.S. Environmental Protection
   Agency
Washington, DC

Dr. Monte A. Mayes
The Dow Chemical Company
Midland, MI

Dr. Judy L. Meyer
University of Georgia
Athens, GA

Mr. Stanton S. Miller
Environmental Science and
   Technology
American Chemical Society
Washington, DC

Dr. Franklin E. Mirer
United Auto Workers
Detroit, MI

Dr. Gerald J. Niemi
Natural Resources Research
   Institute
Duluth, MN

Dr. Ian Nisbet
I.C.T. Nisbet & Company, Inc.
Lincoln, MA

Dr. D. Warner North
Decision Focus, Inc.
Los Altos, CA

Ms. Susan B. Norton
U.S. Environmental Protection
   Agency
Washington, DC

Mr. Richard Orr
U.S. Department of Agriculture
Hyattville, MD

Dr. Richard Parry, Jr.
U.S. Department of Agriculture
Beltsville, MD

Dr. Dorothy Patton
U.S. Environmental Protection
Agency
Washington, DC

Dr. Dennis Paustenbach
McLaren/Hart
Alemeda, CA

Dr. Mary Burr Paxton
ENVIRON
Arlington, VA

Dr. Randall M. Peterman
Simon Fraser University
Burnaby, British Columbia
Canada

Dr. Kieth Pierson
E.I. du Pont Company
Newark, DE

Dr. David Policansky
National Research Council
Washington, DC

Dr. Donald B. Porcella
Electric Power Research
Institute
Palo Alto, CA

Dr. David E. Reichle
Oak Ridge National Laboratory
Oak Ridge, TN

Dr. James Reisa
National Research Council
Washington, DC

Mr. Donald J. Rodier
U.S. Environmental Protection
Agency
Washington, DC

Dr. Andrew Rosenberg
National Marine Fisheries
Service
Woods Hole, MA

Dr. Matthew H. Royer
U.S. Department of Agriculture
Hyattsville, MD

Dr. Marvin A. Schneiderman
National Research Council
Washington, DC

Mr. Peter F. Seligman
Naval Ocean Systems Center
San Diego, CA

Ms. Judith Shaw
American Petroleum Institute
Washington, DC

Dr. Michael Slimak
U.S. Environmental Protection
Agency
Washington, DC

Dr. Eric P. Smith
Virginia Polytechnic Institute and
State University
Blacksburg, VA

Dr. Catherine St. Hilaire
Hershey Foods Corporation
Hershey, PA

Ms. Anne Sprague
National Research Council
Washington, DC

Dr. Ralph G. Stahl, Jr.
E.I. du Pont Company
Newark, DE

Dr. Glenn Suter
Oak Ridge National Laboratory
Oak Ridge, TN

Dr. William H. van der
Schalie
U.S. Environmental Protection
Agency
Washington, DC

Dr. Robert I. Van Hook
Oak Ridge National Laboratory
Oak Ridge, TN

Ms. Joyce Walz
National Research Council
Washington, DC

Dr. Daniel Wartenberg
Robert Wood Johnson Medical
School
Piscataway, NJ

Mr. Michael T. Werner
U.S. Department of Agriculture
Hyattsville, MD

Dr. Bill Williams
U.S. Environmental Protection
Agency
Corvallis, OR

Dr. William P. Wood
U.S. Environmental Protection
Agency
Washington, DC

Mr. Terry Yosie
American Petroleum Institute
Washington, DC

Appendix B

Workshop Organizing Subcommittee and Federal Liaison Group

Lawrence Barnthouse
(Chairman)
Oak Ridge National Laboratory
Oak Ridge, TN

Dr. Douglas Buffington
U.S. Fish and Wildlife Service
Washington, DC

Dr. Robert J. Huggett
Virginia Institute of Marine
  Science
School of Marine Science of the
  College of William and Mary
Gloucester Point, VA

Dr. Ronald Kendall
Institute of Wildlife and
  Environmental Toxicology
Clemson University
Clemson, SC

Dr. Orie L. Loucks
Miami University
Oxford, OH

Alan Maki
Exxon Corporation
Houston, TX

D. Warner North
Decision Focus, Inc.
Mountain View, CA

Ian Nisbet
I.C.T. Nisbet & Company, Inc.
Lincoln, MA

Federal Liaison Group

Dr. Richard Parry, Jr.
U.S. Department of Agriculture
Beltsville, MD

Dr. Michael Slimak
U.S. Environmental Protection
   Agency
Washington, DC

Dr. Maurice Zeeman
U.S. Environmental Protection
   Agency
Washington, DC

Appendix C

Workshop Introduction

One of the principal objectives of the Committee on Risk Assessment Methodology (CRAM) is to determine how risk assessment can be applied to ecological end points. The major environmental problems of the 1990s include such diverse stresses as contamination with toxic substances, overharvesting, habitat destruction, and climate change. Characteristic spatial scales for different types of stresses range from the local to the global. Yet, because priorities must be set at both the national and the local levels, consistent methods are needed for quantifying magnitudes of risks, comparing risks, and making risk-benefit tradeoffs.

A committee was established to plan a workshop on ecological risk assessment. A meeting was held in July 1990 to identify workshop objectives and develop a program. The planning committee agreed that the workshop should survey existing approaches to ecological risk assessment through discussion of specific case studies representative of the major types of environmental stresses, evaluate the applicability of the 1983 four-part risk assessment scheme to environmental assessment and regulation, and identify technical approaches and uncertainties that are common to many environmental problems.

The program began with the three keynote speakers: Terry Yosie, vice-president of the American Petroleum Institute, Mike Slimak, the deputy director of the Office of Ecological Processes and Effects Research, U.S. Environmental Protection Agency, and Warner North, a member of the committee that produced the NRC's 1983 report on human health risk assessment.

Case study presentations followed. Six case-study papers were commissioned for the workshop to provide distinct examples of risk assessment problems: assessing the effects of tributyltin on Chesapeake Bay shellfish populations, testing of agricultural chemicals for effects on avian species, predicting the fate and effects of polychlorinated biphenyls (PCBs) and 2,3,7,8-tetrachlorodibenzo-*p*-dioxin (TCDD) in aquatic ecosystems, quantifying the responses of northern spotted owl populations to habitat change, regulating species introductions, and determining the risks associated with over-harvesting of the Georges Bank multispecies fishery. Each case-study presentation was accompanied by comments from two discussants. The case studies were complemented by eight focused breakout sessions. Four of the breakout sessions were organized around components of the 1983 health risk assessment framework: *hazard identification, dose-response assessment, exposure assessment, and risk characterization.* Each of these sessions was co-chaired by an ecologist and a health risk assessment expert. The purpose of this format was to encourage interaction between the two disciplines and to investigate the applicability of the general concepts developed in the 1983 report to ecological risk assessment. The remaining four breakout sessions were organized around general risk assessment themes: modeling, uncertainty, valuation, and the role of risk assessment in the regulatory process. Dr. Thomas Lovejoy of the Smithsonian Institution was invited to give a closing presentation summarizing his views, based on attendance at the workshop, on the current status and future prospects of ecological risk assessment. Lovejoy's presentation is included in Appendix E.

Participants in the workshop included experts on the specific environmental problems covered in the case studies, representatives of federal and state agencies responsible for performing or evaluating ecological risk assessments, and experts on the technical disciplines (e.g., statistics, ecology, environmental chemistry, and resource economics) that form the scientific basis of ecological risk assessment. The case-study papers, summaries of the discussions and plenary presentations, and the workshop findings are presented in this report.

Appendix D

Opening Plenary Presentations

TERRY F. YOSIE
BUILDING ECOLOGICAL RISK ASSESSMENT
AS A POLICY TOOL

Terry F. Yosie, vice president for health and environment of the American Petroleum Institute and former director of the Science Advisory Board, EPA, provided a broad policy view of issues related to ecological risk assessment. Dr. Yosie noted first that ecological risks have only recently been placed on the nation's policy agenda, in response to increasing public awareness of acid deposition, ozone depletion, climatic change, and other real or potential ecological problems. National and international concern for the environment will stimulate the search for methods and tools for managing ecological risks in the same way that concerns over environmental sources of cancer have stimulated the development of scientific methods and policy tools for regulating human exposure to carcinogens.

Dr. Yosie then posed six questions that must be answered as part of the development of ecological risk assessment methods.

- What is ecological risk assessment?
- Why is it needed?
- What are the key methodological issues in using ecological risk assessment as a policy tool?

• What appropriate lessons can be learned from the health risk assessment experience?
• How should ecological risk assessment be applied?
• What are the needs and future directions for ecological risk assessment?

On the first question, Dr. Yosie noted that previous definitions of ecological risk assessment have ranged from simple statements of principle, such as the definition proposed by the Society for Environmental Toxicology and Chemistry (SETAC, 1987),

a set of formal scientific methods for estimating the probabilities and magnitudes of undesired effects [on plants, animals, and ecosystems] resulting from the release of chemicals, other human actions, or natural catastrophes.

to elaborate schemes for tiered toxicity testing, such as the hazard evaluation procedure used by the Monsanto Corporation (Kimerle et al., 1978). He suggested that an intermediate level of complexity is needed so that the risk assessment methods are simple and accessible enough for writers, policy analysts, and journalists to use with ease, but also technical enough to be useful to scientists and be responsive to advances in science.

Dr. Yosie provided three reasons why policy makers need ecological risk assessment. First, ecological risk assessments can help policy makers correctly diagnose environmental problems before the problems become crises. As an example, he cited EPA's assessment of the health and ecological risks of stratospheric ozone depletion, which stimulated the development of an international agreement to phase out chlorinated fluorocarbons (CFCs). Second, risk assessment is needed to set priorities, as was recently done in the EPA Science Advisory Board Report *Reducing Risk* (EPA, 1990). Third, risk assessment is needed to delineate the link between rational choices and societal values. The debate over global climate change, for example, is ultimately a debate over the responsibility of the current generation to future generations. Risk assessment can clarify the debate by making explicit the climate-change consequences of different policy proposals.

As to the methodological issues central to using ecological risk assessment as a policy tool, Dr. Yosie identified the determination of baseline conditions of ecosystems as a means of assessing the need for protection from development and the estimation of the magnitude of naturally-occurring environmental change in the absence of human intervention.

Dr. Yosie noted that, despite some successes, health risk assessment had yielded few lessons in health-policy decision-making. In part, the failure reflects that it has been more difficult than anticipated to develop and apply health risk assessment methods. In addition, health risk assessments have often been colored by ideological considerations (e.g., arguments over the concept of the maximum exposed individual) or have bogged down over technical questions (e.g., the relevance of rodent data to human risk). In Dr. Yosie's view, these debates have not assisted policy-makers in making informed public health decisions, but they can and they should. There will be increased pressure from EPA, Congress, and the private sector to address broader questions. There is a danger that ecological risk assessment will follow a similarly narrow path, but there is still time to prevent this from happening.

Dr. Yosie provided two examples of how ecological risk assessment can be applied in policy making. Both government and industry, for different reasons, need ecological risk assessment as an aid in contingency planning for and response to oil spills and other kinds of accidents that have ecological impacts. The U.S. oil industry alone will spend more than $900 million over the next 5 years in improving its capability to prevent or respond to oil spills; ecological risk assessment can aid in ensuring cost effectiveness. Similarly, ecological risk assessment can contribute to implementation of total quality management, which is being adopted in many organizations; pollutant releases or obvious ecological impacts can be indicators of inefficient operation.

Finally, Dr. Yosie tried to map out the future of ecological risk assessment with a conceptual policy-triangle. One part of the triangle is ecological risk assessment, which identifies ecological problems, assesses their magnitudes, and provides a perspective on priorities. The second part of the triangle is pollution prevention, aided by the insights provided by ecological risk assessment. Ecological risk assessment and pollution prevention support the third leg, sustainable development. Sustainable development assumes a balanced approach whereby economic growth and environmental improvements proceed together to improve both living standards and the quality of human life and ecosystems.

That approach is already being adopted, as evidenced by the phaseout of CFCs, reduction of chlorine bleaching agents in the manufacture of paper products, and the Administration's recent proposal for "debt for nature" swaps with developing nations.

In conclusion, Dr. Yosie stated that we should develop a capability for ecological risk assessment that is forward-looking and prevention-oriented, as well as backward-looking and remediation-oriented.

D. WARNER NORTH:
RELATIONSHIP OF WORKSHOP TO
NRC'S 1983 RED BOOK REPORT

D. Warner North, a member of the committee that produced the 1983 NRC report *Risk Assessment in the Federal Government: Managing the Process*, provided an overview of the purpose and potential of risk assessment as portrayed in that report. Dr. North argued that the descriptions of the principles of risk assessment and the process for carrying out risk assessment in the 1983 report, which has had a substantial impact on the conduct of human-health risk assessment, provide lessons and insights that apply to ecological risk assessment.

In Dr. North's view, the purpose of the 1983 committee effort was not to provide a summary of risk assessment, but to seek institutional mechanisms for carrying out risk assessment that would be effective in supporting contentious regulatory decisions. The committee found the basic problem in human-health risk assessment to be incompleteness of data—a finding that clearly applies to ecological risk assessment as well. That problem is resolved, not by altering institutional arrangements for performing risk assessment, but by improving the process by which risk assessments are made.

Perhaps the most widely reproduced part of the 1983 report is its description of the elements of risk assessment and risk management. These are reproduced in the current committee report as Figure 3-1 (Chapter 3). The elements collectively provide a bridge between science and risk management, which might be more generally denoted as policy. Risk assessment can provide a consistent process for summarizing science to support regulatory decision-making by federal agencies. The process of providing the scientific basis must be consistent and flexible before regulatory policies for managing risks can be evolved that are consistent and yet permit change based on the evolution of scientific knowledge.

Scientific knowledge is incomplete, and the multiplicity of resulting uncertainties needs to be dealt with by making choices among sets of scientifically plausible options. Rather than having those critical choices left to the discretion of individual risk assessors or the influence of risk managers, who might desire to regulate or not regulate in a specific situation on the basis of nonscientific considerations, the 1983 report suggests these choices can be made systematically with a risk assessment policy that is consistent with science and permits exceptions based on science. The report presents as its lead recommendation that

> regulatory agencies should maintain a clear conceptual distinction between assessment of risks and the consideration of risk management alternatives; that is, the scientific findings and policy judgments embodied in risk assessments should be explicitly distinguished from the political, economic, and technical considerations that influence the design and choice of regulatory strategies.

The implication is that the scientific issues resulting from gaps in data and in theoretical understanding should be dealt with in a consistent and predictable way. Furthermore, the scientific issues should be carefully distinguished from nonscientific issues on which policy discretion is expected and for which the decision-maker is held responsible.

It can be argued that health risk assessment practice has gone too far in separating risk assessment and risk management. The 1983 report advocated conceptual distinction, not separation. The report states that

> the importance of distinguishing between risk assessment and risk management does not imply that they should be isolated from each other; in practice they interact, and communication in both directions is desirable and should not be disrupted.

Furthermore, risk assessment must serve an assortment of functions in support of risk management, from initial screening and priority-setting exercises to major regulatory decisions with profound economic and public health consequences. Simple procedures appropriate for screening and priority-setting "may have to yield to more sophisticated and detailed scientific arguments when a substance's commercial life is at stake and the agency's decision may be challenged in court." Unfortunately, the same simple procedures used for health risk assessment in simple screening applications have often been used for risk assessment in support of major regulatory decisions as well. Rarely has the flexibility been used to bring in "more sophisticated and detailed scientific arguments" to replace the default assumptions, even though such departures are permitted under health risk assessment guidelines.

Dr. North then returned to a discussion of the four elements or steps in the risk assessment paradigm. Not all these steps are always required: a risk assessment might stop with the first step, hazard identification. The definitions of the steps can be translated from the context of health risk to the context of ecological risk quite readily:

- *Hazard identification.* The determination of whether a particular chemical (stress agent) is or is not causally linked to particular ecological effects.
- *Dose-response assessment.* The determination of the relation between the magnitude of exposure and the probability of occurrence of the effects in question.
- *Exposure assessment.* The determination of the extent of exposure before or after application of regulatory controls.
- *Risk characterization.* The description of the nature and often the magnitude of ecological risk, including attendant uncertainty.

However, the terms *magnitude of exposure* and *extent of exposure* might require replacement with a more general measure of ecological stress.

The major thrust of the 1983 report was not to recommend that risk assessment be carried out with the four steps. Rather, most of the recommendations addressed the process of summarizing the science in support of risk management. In addition to the first recommendation on the conceptual distinction between risk assessment and risk management, the report recommended that risk assessments be made publicly available as written documents in advance of regulatory decisions and that such risk assessments be subjected to peer review by scientists from outside the agency. Uniform guidelines should be developed for the use of federal agencies in the risk assessment process, and these guidelines should be comprehensive, detailed, and flexible enough "to consider unique scientific evidence in particular instances." Finally, the 1983 report recommended establishment of a "Board on Risk Assessment Methods." Some of the suggested functions of the board are now being carried out by CRAM.

MICHAEL SLIMAK:
U.S. ENVIRONMENTAL PROTECTION AGENCY
ACTIVITIES IN ECOLOGICAL RISK ASSESSMENT

Michael Slimak, deputy director of EPA's Office of Ecological Processes and Effects Research, presented an overview of EPA's past and present activities in ecological risk assessment. Dr. Slimak identified five major problems that have made these assessments difficult to perform in a consistent way:

• The need to consider multiple species and levels of biological organization;
• The diversity and multiplicity of end points (e.g., mortality and biochemical cycling);
• The simultaneous actions of multiple stressors, such as pollution and habitat loss;
• The difficulty of relating ecological changes to societal values;
• The multiplicity of regulatory mandates under which EPA operates.

Dr. Slimak defined ecological risk assessment as a "probabilistic statement of the 'outcome' [effects] associated with an ecological receptor being exposed to some form of stress." He then described some of the agency's approaches to assessing exposures and outcomes, focusing on two generic classifications: predictive or "bottom-up" assessments for single chemicals, as exemplified by the regulation of pesticides and toxic chemicals, and holistic or "top-down" assessments, such as assessments of wetland loss, effects of acid deposition, and global climate change. Most of EPA's attention has been devoted to predicting ecological effects of single chemicals from laboratory toxicity-test data. Although relatively elaborate guidelines and procedures have been developed for this purpose, the predictive approach has inherent weaknesses that have long been recognized. Recently, water-quality regulation has moved toward a top-down approach based on measurement of community integrity from field data.

Many of the problems facing EPA are not amenable to the predictive approach, either because they involve stresses other than toxic chemicals or because they involve direct observation of adverse ecological changes. Examples discussed by Dr. Slimak include explanation of dolphin-stranding incidents, a reported worldwide amphibian decline, and performance of ecological assessments at Superfund sites. Such studies involve difficult scientific problems. The National Acidic Precipitation Assessment Program's assessment of the relationship of sulfur dioxide deposition to aquatic resource quality (NAPAP, 1991) best demonstrates problems encountered by EPA.

For the last 5 years, EPA has been conducting an ecological risk assessment research program focused on developing better predictive models for single-chemical assessments. A major new initiative, the Environmental Monitoring and Assessment Program (EMAP), will attempt to measure ecosystem quality on regional and national scales through a nationwide monitoring program. The results will be used to determine the success of EPA's regulatory programs and to support future risk assessments.

The EPA Risk Assessment Forum has initiated the development of guidelines for ecological risk assessment analogous to the existing guidelines for health risk assessment. A series of risk assessment colloquia was held during 1990. The proceedings were summarized and published in early 1991 in a report entitled *Issues in Ecological Risk Assessment.* A "framework document," intended to provide the conceptual basis for detailed guidelines, is now being reviewed (EPA, 1992a). A strategy for subject-specific guidelines structured around ecosystem types, levels of biological organization, end points, and stressors is being developed simultaneously. Case studies illustrating current practice are being developed and a report containing case studies will be announced in the *Federal Register* (EPA, 1992b).

Dr. Slimak closed his presentation by raising issues for consideration at the CRAM workshop:

- The amenability of ecological risk assessments to biostatistical treatment;
- End point identification and selection;

- Ecological values;
- The relationship of the 1983 paradigm to regulatory processes;
- The relationship between risk assessment and risk management.

Appendix E

Case Studies and Commentaries

CASE STUDY 1:
Tributyltin Risk Management
In the United States
R. J. Huggett and M. A. Unger,
Virginia Institute of Marine Sciences

Tributyltin (TBT) is a chemical with a variety of biocidal applications, including use as an antifouling agent in boat paints (Blunden and Chapman, 1982). Biological effects of TBT on marine and estuarine organisms and the concentrations of TBT that induce them vary widely among species (Huggett et al., 1992). A water concentration of 1,000 ng/L (1 part per billion) is lethal to larvae of some species, and nonlethal effects have been observed at concentrations as low as 2 ng/L (2 parts per trillion, ppt). Both laboratory and field studies of toxicity were initially hampered by difficulties in measuring the low concentrations that were toxic to some organisms.

Adverse effects on nontarget organisms, including commercially valuable species of shellfish, were observed in Europe in the early 1980s (Alzieu, 1986; Abel et al., 1986). Abnormal shell growth was documented in *Crassostrea gigas* (European oyster) and linked through laboratory experiments to TBT leached from antifouling paints. That connection led to restrictive regulations in France (in 1982) and Great Britain (in 1985 and 1987). In the United States, concentrations exceeding those determined experimentally to be effective have been found in many areas, particularly in harbors with large marinas. Snails in the vicinity of a marina on the York River, Virginia, were shown to have an abnormally high incidence of imposex (expression of male characteristics by female organisms), an effect previously observed under laboratory conditions in female European oysters, *Ostrea edulis* (Huggett et al., 1992). EPA began to assess effects of TBT in 1986, but has not yet issued any regulations. Meanwhile, restrictive actions have been taken by states and by the Congress.

A proposal by the U.S. Navy to use TBT paints on its entire fleet was prohibited by Congress in 1986, despite a Navy study that predicted no adverse environmental impact. Virginia enacted legislation and an emergency regulation in 1987, and Maryland, Michigan, and other states have since taken similar actions. Congress enacted national legislation restricting use of TBT paints in 1988. Those actions generally banned or restricted the use of TBT paints on small boats (less than 25 m long) and placed limits on leaching rates from paints used on larger vessels. Studies in Virginia had shown that most TBT releases were from small boats. Small-scale monitoring studies (e.g., in France and Virginia) have shown that the restrictions have been effective in reducing environmental concentrations and adverse impacts of TBT.

Risk management of TBT has been unusual in several ways. The initial basis for concern was field observation of adverse effects, not extrapolation from laboratory bioassays and field chemistry data. Risk assessment and risk management were conducted by state agencies and legislatures, rather than by EPA. Although the risk assessments were made without formalized methods, the results of the independent assessments were the same. Finally, TBT is the first compound banned by the Congress and the first regulated for environmental reasons alone.

Discussion
(Led by L. Barnthouse, Oak Ridge National Laboratory, and P. F. Seligman, Naval Ocean Systems Center)

The case study addressed, with differing completeness, each of the five recommended steps in risk assessment and management. Hazard identification included the observation of abnormalities in the field and the same effects in experimentally exposed animals. Dose-response identification included data both from the field (correlative) and from the laboratory (experimental). Exposure assessment was based on estimated use and release rates rather than on monitoring or modeling studies. Risk characterization was only qualitative; it did not address such issues as the number and distribution of species that were vulnerable, or the degree of damage to the shellfish industry. Risk management actions were based on the demonstrable existence of hazard, on societal concern for the vulnerable species, and on the ready availability of alternative antifouling agents.

Some workshop participants were critical of the risk assessment approach adopted by Congress and state regulatory agencies. No attempt was made to plan and execute a formal risk assessment. Risk identification was based primarily on data on nonnative species. The Eastern oyster and blue crab, the species putatively at greatest risk, have been found to be less sensitive. Regulatory responses were based on findings of high environmental concentrations of TBT in yacht harbors and marinas, rather than in ecologically important regions such as breeding grounds. The central issue is whether a safe loading capacity (environmental concentration) of TBT for nontarget organisms can be defined, given substantially reduced rates of input. Recent information on fate and persistence, chronic toxicity, and dose-response relationships could support a more quantitative risk assessment with the possibility of more or less stringent restrictions.

CASE STUDY 2:
Ecological Risk Assessment for Terrestrial Wildlife
Exposed to Agricultural Chemicals
R. J. Kendall, Clemson University

The science of ecological risk assessment for exposure of terrestrial wildlife to agricultural chemicals has advanced rapidly during the 1980s. EPA requires detailed assessments of the toxicity and environmental fate of chemicals proposed for agricultural use (EPA, 1982; Fite et al., 1988). Performance of an ecological risk assessment requires data from several disciplines: analytical toxicology, environmental chemistry, biochemical toxicology, ecotoxicology, and wildlife ecology.

Addressing the ecological risks associated with the use of an agricultural chemical involves a complex array of laboratory and field studies—in essence, a research program. This paper provides examples of integrated field and laboratory research programs, such as The Institute for Wildlife and Environmental Toxicology (TIWET) at Clemson University. Preliminary toxicological and biochemical evaluations include measurements of acute toxicity (LC_{50} and LD_{50}), toxicokinetics, and observations of wildlife in areas of field trials. Assessment of reproductive toxicity includes studies with various birds and other wildlife, particularly European starlings that nest at high densities in established nest boxes; these studies include measurements of embryo and nestling survival, postfledgling survival, behavior, diet, and residue chemistry (Kendall et al., 1989). Nonlethal assessment methods include measurement of plasma cholinesterase activity associated with organophosphate pesticide exposures (Hooper et al., 1989). A wide variety of birds, mammals, and invertebrates have been used in these studies.

End points evaluated in wildlife toxicological studies include mortality, reproductive success, physiological and biochemical changes, enzyme impacts, immunological impairment, hormonal changes, mutagenesis and carcinogenesis, behavioral changes, and residues of parent compounds and metabolites (Kendall, 1992).

The paper includes a case history of a comparative evaluation of Carbofuran and Terbufos as granular insecticides for control of corn rootworms. Carbofuran has been responsible for many incidents of wildlife poisoning and is recognized as being very hazardous to wildlife. In contrast, although Terbufos is highly toxic to wildlife in laboratory studies, exposure of wildlife under field conditions appears generally to be relatively low, and widespread mortality is not evident. Field studies of Terbufos conducted by TIWET might be the only ones conducted to date that satisfy EPA's requirements for a Level 2 field study, a more quantitative assessment of the magnitude of the effects of a pesticide than the qualitative Level 1 studies. (Level 2 studies are performed when toxicity tests and use patterns suggest a detailed study is warranted.) Data generated in those studies support an ecological risk assessment for Terbufos that is reported in the paper. However, the research program on Terbufos represents many years of effort with integration of laboratory and field research to achieve a full-scale level 2 study in just one geographic area on one crop. Ecological modeling techniques will be needed to generalize the results to other chemicals or to other situations.

Discussion
(Led by B. Williams, Ecological Planning and Toxicology, Inc., and J. Gagne, American Cyanamid Company)

Dr. Williams noted that each step in ecological risk assessment is more complex and less understood than the corresponding step in human-health risk assessment. Although hazard can be assumed when a toxic chemical is released, the species and populations at risk must first be defined. The appropriate selection of surrogate species for testing in the laboratory is usually unclear. Measurement of environmental concentrations is only the first step in exposure characterization. Exposure assessment also requires consideration of foraging behavior, avoidance, and food-web considerations, as well as spatial and temporal variability. Risk characterization involves comparison of exposure estimates with measures of hazard; this process might result in compounding of errors. Ecological risk assessments do not track individuals over time and so do not accurately reflect population changes.

The activities presented in the case study have a large research component, which is focused on dose-response assessment and exposure assessment. One discussant characterized risk assessment, as presented in the case study, as a retrospective exercise based on focused characterization of hazard and exposure in wildlife. Given the difficulties in conducting environmental risk assessments, the four-part paradigm might not be applicable at levels of organization above that of the population.

CASE STUDY 3A:
Models of Toxic Chemicals in the Great Lakes: Structure, Applications, and Uncertainty Analysis
D. M.DiToro, Hydroqual, Inc.

This paper reviewed and summarized efforts to model the distribution and dynamics of toxic chemicals in the Great Lakes, with applications to PCBs, TCDD, and other persistent, bioaccumulated compounds. The models were based on the principle of conservation of mass (Thomann and Di Toro, 1983). Analysis proceeded through five steps: water transport, dynamics of solids, dynamics of a tracer, dynamics of the toxicant, and bioaccumulation in aquatic organisms. Mechanisms considered include settling, resuspension, sedimentation, partitioning, photolysis, volatilization, biodegradation, growth, respiration, predation, assimilation, excretion, and metabolism. The model of toxicant dynamics considered three phases (sorbed, bound, and dissolved) in each of two media (water column and sediments) and 21 pathways into, out of, or between these phases. The model of bioaccumulation included 25 compartments (four trophic levels with one to 13 age classes at each level) with five pathways into or out of each compartment. Because of the large number of coefficients (rate constants), sparseness of knowledge of inputs, and little opportunity for field calibration, uncertainty analysis was important in all the modeling exercises.

The first example modeled the dynamics of total PCBs in Lake Michigan (Thomann and Connolly, 1983). Plutonium-239 was used as a tracer to analyze sediment dynamics, and the model suggested that resuspension is an important mechanism. Calculation of PCB concentrations was limited by an order-of-magnitude uncertainty in the mass loading. Predictions of PCB concentrations and their rate of decline were sensitive to the value assumed for the mass-transfer coefficient for volatilization.

The second example modeled TCDD in Lake Ontario and attempted to predict the relationship between one source of input and the resulting incremental concentrations of TCDD (Endicott et al., 1989). In the absence of knowledge of other inputs, field data could not be used to calibrate the model. Hence, a formal uncertainty analysis was performed with Monte Carlo methods and assumed probability distributions of the rate coefficients. The 95% confidence limits of predicted TCDD concentrations in water and sediment differed by a factor of 10-100. Uncertainties in rate constants for photolysis and volatilization were the most important sources of uncertainty in predicted TCDD concentrations.

The third example extended the Lake Ontario TCDD model to eight other hydrophobic chemicals and incorporated a food-chain model to predict concentrations in lake trout (Endicott et al., 1990). The model predicted wide differences in toxicant concentrations, depending primarily on the degree of hydrophobicity as indexed by the octanol-water partition coefficient, Kow. The range of uncertainty in the predicted concentrations also varied among the chemicals. In-lake removal processes (sedimentation, volatilization, and degradation) were important for all chemicals.

CASE STUDY 3B:
Ecological Risk Assessment of TCDD and TCDF
M. Zeeman, U.S. Environmental Protection Agency

This paper is based on a full-scale ecological risk assessment of chlorinated dioxin and furan emissions from paper and pulp mills that use the chlorine bleaching processes (Schweer and Jennings, 1990). Although the risk assessment addressed potential risks to terrestrial and aquatic wildlife exposed to TCDD and 2,3,7,8-tetrachlorodibenzofuran (TCDF) via a number of environmental pathways, the case study was limited to exposure of terrestrial wildlife to TCDD resulting from land disposal of paper and pulp sludges. This route of exposure was identified as one of the most hazardous in the multiroute risk assessment.

The specific exposure pathway considered was uptake of TCDD by soil organisms (earthworms and insects) from soil to which pulp sludge has been applied, and the consumption of soil organisms by birds and other small animals. Transfer factors were estimated both by modeling and from data collected in a field study in Wisconsin, in which an average soil TCDD concentration of 11 ppt led to concentrations of up to 140 ppt in a composite of six robin eggs. The models used three alternative sets of assumptions: low estimate, best estimate, and high estimate. The best estimates of tissue concentrations derived from the model were often similar to those observed in the field study: the low and high estimates were lower and higher, respectively, by a factor of roughly 10.

Risk estimates for terrestrial wildlife were derived by comparing exposure estimates (usually converted to daily intake rates) with benchmark toxicity values. The values used as benchmarks were either lowest-observed-adverse-effect levels (LOAELs) or no-observed-adverse-effect levels (NOAELs) for reproductive toxicity in birds and mammals

—specifically, the lowest reported LOAELs and NOAELs. The risk quotient (RQ) for each species considered was defined as the ratio of the estimate of exposure to the corresponding benchmark value. On the basis of transfer estimates for land disposal of paper sludges, RQs could exceed 60:1 for the most exposed species (robins, woodcocks, and shrews). To estimate soil concentrations of TCDD "safe" for these species, two uncertainty factors of 10 could be applied: one to allow for interspecies variability in sensitivity and one for an extrapolation from laboratory to field and/or the use of a LOAEL as the benchmark value. The corresponding estimates of safe concentrations were estimates that would lead to RQs less than 0.01:1 for the most heavily exposed species considered. Under those assumptions, soil concentrations of TCDD safe for highly exposed species would be about 0.03 ppt.

Discussion
(Led by L. A. Burns, U.S. Environmental Protection Agency, and D. J. Paustenbach, McLaren/Hart)

These case studies present only estimates of environmental concentrations—i.e., exposure assessment—and do not address other elements of risk assessment. Compared with traditional human-health assessments, they show a greater concern for accuracy (as opposed "policy-driven conservatism"), a greater use of formal uncertainty analysis, and better opportunities for verifying accuracy of exposure and uptake models.

Criticism of the models focused on the omission of processes and on the assumed linear relationship between loading and environmental concentrations. Omitted processes include in-lake generation of solids (phytoplankton), transport in the benthic boundary layer, effects of water clarity on photolysis rates, and daily cycles in pH. A nonlinear relationship between loading and toxicant concentrations might occur if the toxicant reaches high enough concentrations to change the processes that control its own fate. For example, reduction in fish populations might allow for higher populations of zooplankton, which clarify the water column by decreasing populations of phytoplankton, thereby increasing photolysis rates and stabilizing pH.

CASE STUDY 4:
Risk Assessment Methods in Animal Populations:
The Northern Spotted Owl as an Example
D. R. Anderson, U.S. Fish and Wildlife Service

This paper described an analysis of northern spotted owl population dynamics performed to support ongoing studies of the impacts of clearcutting of old-growth forest on the prospects for future survival of this endangered species (Salwasser, 1986). The paper summarized a method for estimating rates of population increase or decrease based on capture-recapture techniques and illustrates the methods with data on the northern spotted owl. The method proceeds in three steps: use of capture-recapture data to estimate age-specific survival or fecundity rates, estimation of the finite rate of population change (Leslie's parameter λ), and experiments on samples of marked animals in natural environments. Mathematical models for estimating population parameters, including λ, have been developed extensively, and computer programs are available (Burnham et al., 1987). Experimental studies are desirable to test hypotheses about relationships between population parameters and risk factors.

The case study was of a population of northern spotted owls in California studied for 6 years (Franklin et al., 1990). Capture-recapture data yielded estimates of age-specific survival and fecundity for females, as well as estimates of mean population size (37 females) and annual recruitment (0 to 19 females; mean, 8). On the average, the eight females entering the population each year would have included six immigrants from outside the study area and only two locally raised recruits. The calculated value of λ was 0.952 ± 0.028, which indicated a decreasing population.

In this case, the risk factor was clearance of the old-growth forest on which the species is believed to depend. Although the study area contained much suitable habitat, the population appeared not to be self-sustaining, but to be maintained by immigration from remaining areas of old growth. It was suggested that the study population is temporarily above the long-term carrying capacity because of the drastic loss of habitat in surrounding areas; these circumstances lead to a large "floating" component of the population.

The paper concluded that risk assessment in higher vertebrate populations must often rely on analysis of samples of marked individuals. A robust theory exists for study design and the analysis of such data. Selection of appropriate models is critical for rigorous assessment of impacts. Analysis of capture-recapture data allows inferences about the separate processes of birth, death, emigration, and immigration. Risk to a population does not affect population size directly; rather, it acts on the fundamental processes of birth and death.

Discussion
(Led by M. E. Kentula, U.S. Environmental Protection Agency, and O. L. Loucks, Miami University)

Dr. Kentula commented that the case study (like others in the workshop) focused on individuals and populations and thus took a bottom-up approach. An alternative, top-down approach is to conduct an ecosystem risk assessment from a landscape perspective. For example, Kentula stated that EPA's Wetlands Research Program is developing methods to assess impacts on landscape function due to cumulative wetlands loss (Abbruzzese et al., 1990). The method proceeds in two stages: a landscape characterization map is used to classify and rank units of the landscape according to relative risk, and can also be used to set priorities for effort and allocation of resources; a response curve expresses the hypothesized relationship between stressors (such as loss or modification of wetlands) and reduction in landscape functions (e.g., maintenance of water quality, or life support). The system can be used both to identify areas at risk and to guide management decisions for landscapes that are already affected.

Dr. Loucks commented that the case study presents the consequences of the stress to one local owl population at one time. For assessment of risk to the regional or total population, one would need to construct a "dose-response" relationship, in which "dose" would be a measure of the degree of stress (e.g., the percentage of the old-growth forest that has been destroyed) and "response" would be the probability of extinction of the population within an appropriate period (e.g., 250 years). Calculation of the probability from the birth, death, and dispersal rates estimated in the case study would require stochastic population modeling that takes account of uncertainty and variability in the population parameters. The Endangered Species Act is an example of preemptive risk management, in that a high probability of extinction of a single species is designated as unacceptable. A species-by-species approach, however, does not lead to quantitative assessment of the risk of impoverishment of an ecosystem. Where possible, ecological risk assessment should work across levels of organization and should assess risks of reduction in system utility.

CASE STUDY 5:
Ecological Benefits and Risks Associated with the Introduction of Exotic Species for Biological Control of Agricultural Pests
R. I. Carruthers, USDA Agricultural Research Service

The accidental or deliberate introduction of exotic species into regions where they are not native can cause positive, negative, or no observable effects, depending on a wide variety of biological, sociological, economic, and other factors. About 40% of the major arthropod pests (Sailer, 1983) and 50-75% of weed species (Foy et al., 1983) in the United States are introduced species, and introduced pests also include vertebrates, mollusks, and disease organisms that affect animals and plants. Many countries have developed formal programs to limit the introduction and establishment of unwanted exotic organisms, and many have developed methods to assess benefits and risks associated with planned introductions. The United States has no federal statute or set of statutes that governs introductions; instead, it has cumbersome and sometimes conflicting regulations, protocols, and guidelines.

This paper addressed assessment of risks and benefits of "classical biological control" (CBC): the planned introduction of exotic enemies of an introduced pest collected from the pest's home range (DeBach, 1974). Classical biological control (either alone or integrated with other pest management methods) has frequently been successful in controlling introduced pests and often provides large economic or environmental advantages over alternative methods. An example given in the paper is control of the alfalfa weevil: introduction and widespread releases of 11 species of parasitic hymenoptera have yielded substantial control of this major pest with no known negative side effects and with an estimated benefit-to-cost ratio of 87:1.

Risks of CBC programs have three different sources: the organism itself (e.g., parasitism or predation on nontarget species), associated organisms (e.g., pests of the introduced beneficial organism), and unrelated passenger organisms arriving with shipments of the introduced organism. Some adverse effects of all three types have been documented (Pimentel et al., 1984, Howarth, 1991), including local extinctions of nontarget species, especially in island situations. Although there is little documentation of notable adverse impacts of CBC programs in the United States, more precise prediction of benefits and risks would be desirable. Unfortunately, accurate prediction of both positive and negative impacts (target and nontarget effects) of CBC programs has not been achieved. The lack of predictive ability leaves CBC risk assessments in the realm of informed scientific judgment based on limited published data.

In addition to requirements of various federal laws, guidelines have been developed to improve safety in CBC. Agricultural Research Service protocols (now under revision) require federal permits for importation and movement of organisms, quarantine, authoritative identifications, environmental and safety evaluations, documentation of movements and releases, and retention of voucher specimens. Current policy requires an environmental assessment (EA) to accompany applications for permits for field release of exotic organisms. Although the components of an EA depend on the specific situation, the documentation required is fairly extensive. At any step in the process, a proposed introduction can be deemed inappropriate and the project terminated.

Discussion
(Led by J. T. Carlton, Williams College, and
D. Policansky, National Research Council)

Classical biological control is only one kind of introduction of nonnative species. Others include range expansions (either natural or mediated by human modification of habitats), deliberate introductions to "improve nature" or for aquaculture or horticulture, and a wide variety of accidental introductions. CBC seems to have a better safety record than other types of introduction. It is not clear whether this is because the activity is basically benign, because the safety precautions work well, or because CBC involves small organisms that pose smaller risks than larger organisms. The worst failures in all categories have occurred in insular environments such as islands and lakes.

The assessment of risks posed by introductions has been addressed separately by scientists in different disciplines (e.g., agriculture, freshwater and marine ecology, and nature conservation). Communication between the disciplines is poor, and several sets of criteria, procedures, and protocols have been developed independently. Whereas the U.S. Department of Agriculture has adopted flow charts as a way to systematize decision-making, other agencies (e.g., the International Council for the Exploration of the Sea) have concluded that too little is known about ecosystem functioning for flow charts to be useful.

Dr. Policansky commented that risk assessment for species introductions is difficult to fit into the four-step Red Book paradigm. Hazard is taken for granted (because it is the introduction of the species itself); dose-response and exposure are yes-no categories, not continuous variables, because the more important point is whether the species is present or not, not how much of the species is present. A more suitable paradigm might be that presented in the 1986 NRC report *Ecological Knowledge and Environmental Problem-Solving: Concepts and Case Studies*, which placed more emphasis on problem-scoping and problem-solving than on categorical activities.

CASE STUDY 6:
Uncertainty and Risk in an Exploited Ecosystem:
A Case Study of Georges Bank
*M. J. Fogarty, A. A. Rosenberg, and M. P. Sissenwine,
National Marine Fisheries Service*

This paper addressed the risks of overexploitation of harvested marine ecosystems, with specific application to Georges Bank, a highly productive area off the northeastern United States. In this context, risk assessment involves determining the probability that a population will be depleted to an arbitrarily predetermined "small" (e.g., 1% or 5%) size. The "quasi-extinction" level may be defined (Ginzburg et al., 1982) as (1) the population level below which the probability of poor recruitment increases appreciably or (2) the smallest population capable of supporting a viable fishery.

The primary determinant of the long-term dynamics of any population is the relationship between the adult population (stock) and recruitment. The null hypothesis is that the relationship is linear, i.e., that recruitment is independent of density (Sissenwine and Shepherd, 1987). Compensatory changes in survival or in reproductive output result in nonlinear stock-recruitment curves. Nonlinearity permits stable equilibrium under harvesting pressure (i.e., under increased mortality rates), up to a critical exploitation level, beyond which the population will decline to quasi-extinction. Stochastic variation in the stock-recruitment relationship or in multispecies interactions can increase risks of adverse effects at moderate exploitation levels. In practice, because of uncertainties resulting from stochastic variations and measurement errors, it is often impossible to reject the null hypothesis of no compensation. Assuming there is no compensation will, in general, result in a conservative assessment of production capacity and its ability to withstand exploitation.

Haddock populations on Georges Bank fluctuated about relatively stable levels between 1930 and 1960 when the fraction of the total haddock population killed per year by fisherman (annual fishing mortality rate) varied between 0.3-0.6, but collapsed after the fishing mortality rate increased to 0.8 during the 1960s (Grosslein et al., 1980). The empirical relationship between stock and recruitment was extremely variable with little indication of the form of the underlying curve. Analysis of the population dynamics showed that a density-independent null model could not be rejected and gave a neutral equivalent harvest rate of 0.5, which agrees well with the stable period of the fishery. In contrast, the compensatory model is over optimistic with respect to the long-term harvest rate.

The decrease in populations of haddock and other groundfish was accompanied by increases in other species, notably elasmobranchs (rays and sharks). The biomass of predatory species increased dramatically with attendant consequences for the overall system structure (Fogarty et al., 1989). Population modeling suggests that the stock-recruitment relationship for haddock might have been changed and that the population cannot now withstand as heavy fishing mortality as it could before the increase in predation pressure.

Risk assessment for exploited systems must take into account uncertainties in population abundance, harvest rates, and system structure. Adoption of risk-averse management strategies would minimize the possibility of stock depletion or undesirable alterations in the structure of the system.

Discussion
(Led by R. M. Peterman, Simon Fraser University, and
J. L. Ludke, National Fisheries Research Center-Leetown)

Discussion focused on the idea of statistical power—the probability that an experiment (or set of observations) will correctly reject a null hypothesis that is false, i.e., the probability that an experiment will detect effects that actually exist. In fisheries cases, the high degree of variability in population parameters means that most studies have very low power to detect changes, unless the studies are continued for many years or involve frequent measurements (Peterman and Bradford, 1987). Published papers in fisheries biology (and in other disciplines related to risk assessment) rarely report statistical power and hence can misleadingly report negative findings. The case study recommended adopting a conservative null hypothesis to allow for the low power of the observational studies. Other approaches are to improve the design of studies (e.g., by more frequent sampling), to incorporate uncertainties into formal decision analysis, and to reverse the burden of proof (to put the burden of documenting whether detrimental effects are occurring on exploiters of the resource, rather than in the management agency). If "proof" of safety is required, a formal statement of the power of studies should be provided for a size of effect deemed relevant.

The Georges Bank fishery is only one of a long series of cases in which overexploitation has occurred despite a nominal system of scientific stock assessment and fishery management. Discussants generally felt that overexploitation was due to failures of management, rather than to deficiencies in assessment or failure to communicate results to managers.

The assessment of the risk to fish populations associated with exploitation in the Georges Bank case study is implicitly consistent with the 1983 health risk assessment framework, although the explicit steps differ. The case study illustrates the 1983 risk assessment paradigm within the larger context of problem-solving. However, the dose-response and exposure steps might be only loosely analogous. Differing circumstances of function, scale, and certitude could require variation in the method of risk assessment.

The numerous sources of uncertainty in assessing risk associated with exploitation of fish populations vary and increase in magnitude with increase in scale. Regulation of harvest of geographically confined populations can be achieved with greater confidence than can regulation of wide-ranging populations such as Chesapeake Bay striped bass and Lake Michigan lake trout. Sources of uncertainty include variation in recruitment, measurement (which requires many assumptions), and management and institutional characteristics. Management techniques for reducing risks associated with overexploitation of populations are fairly blunt instruments, and strong actions are usually taken only after the fact. Rarely, if ever, are risk-reduction measures considered until an actual impact is noticed or a potential threat emerges.

Subtle and cumulative factors that are unknown or are measured imprecisely—e.g., chronic or episodic changes in predation, migration, and disease—are some of the issues with information gaps that contribute to uncertainties in ecological risk assessment. The Georges Bank case study describes multispecies interactions and consequences of selective harvesting practices within the fish community, but falls short of a systematic understanding of cause and effect with regard to changes in multispecies abundance.

Appendix F

Breakout Sessions

HAZARD IDENTIFICATION
A. Maki and D. Patton

The hazard-identification group examined the case studies in light of the 1983 Red Book paradigm and experience with Environmental Protection Agency (EPA) guidelines for health risk assessments to set the context for discussing hazard identification in ecological risk assessment. Generic issues related to paradigm flexibility, scope of ecological risk assessment, the role of uncertainty in research, and the role of nonscientific consideration were discussed. Specific issues were examined for each case study in terms of ecological hazard.

Generic Issues

There was general agreement that flexibility existed (even if not always applied) in the 1983 paradigm and in forthcoming EPA health guidelines. Flexibility is desirable for ecological risk assessment. Although the four components of the paradigm—hazard identification, dose-response assessment, exposure assessment, and risk characterization—are appropriate for any ecological risk paradigm, they may be combined in different ways. For example, hazard identification may be combined with other steps or treated separately case by case. The group also agreed that uncertainties that were not fully analyzed for hazard identification in the case studies are as important in the presentation of hazard data as they are for health risk assessment.

Discussion of other questions suggested that the scope and definition of ecological risk assessment might be broader than the scope and definition of human health risk assessment in the Red Book. For example, risk management considerations (management and political pressures, social costs, economic considerations, and regulatory outcomes) were ingredients in all case studies and related discussions. Much attention was paid to the influence of management on the scope and design of assessment. Such considerations are absent from discussions of health risk assessment. Some participants also felt that generation of new data should be treated as an aspect of risk assessment, rather than restricting *risk assessment* to evaluation of data that are already in hand.

Discussion leaders questioned the role of valuation in hazard identification, but this issue was not discussed in detail. In view of repeated references to the question of end-point selection as a valuation decision, additional examination on this point is needed.

The case studies illustrated the importance of a systematic presentation and evaluation of data used to identify hazard. Discussion leaders noted that presentation of hazard data was highly variable in the case studies and suggested that some of the hazard identification principles that guide health hazard evaluation might be useful, including emphasis on a complete and balanced picture of relevant hazard information. Specific criteria and questions that are critical to identifying ecological risk are needed to develop an operational definition of *complete and balanced.*

Analysis of Case Studies

Examination of the case studies revealed a variety of approaches to ecological hazard identification.

For the tributyltin study, hazard identification was based initially on field studies. Retrospective epidemiological studies included a monitoring program (both biological and chemical) and laboratory investigation of cause-effect relationships.

In pesticide risk assessments, as exemplified by the agricultural-chemicals case study, neither laboratory nor field studies are required to establish a hazard. Instead, there is a regulatory presumption of hazard.

A similar presumption of hazard is used by the U.S. Department of Agriculture in evaluating proposed species introductions for biological control purposes.

The polychlorinated biphenyls and 2,3,7,8-tetrachlorodibenzo-*p*-dioxin study did not explicitly discuss hazard identification. Regulatory actions on both substances are strongly influenced by human health risks, so it is not clear that any explicit ecological hazard identification was needed or performed.

For the spotted owl study, hazard identification occurred through environmental impact studies undertaken by federal agencies to comply with National Environmental Policy Act that identified this species as being vulnerable to loss of habitat due to old-growth forest clearing.

In fisheries management, it might be assumed that fishing is by definition a hazard. Within limits, fishing confers no greater risk to a population than does predation or even the killing of small numbers of fish by toxic chemical spills. Detailed assessments, such as those described in the case study, appear to be triggered by observations of declining catch or by other evidence (e.g., from modeling studies) that suggests that sustainable yields are being exceeded.

The case studies demonstrate that ecological hazard identification can take many forms and can involve both scientific data and policy decisions. The group discussed two possible modifications of the Red Book paradigm to accommodate the clear influence of policy on the conduct of ecological risk assessment: addition of a "scoping" component before hazard identification and expansion of the definition of *hazard identification* to include management inputs. No consensus was achieved on which alternative is preferable, but the group agreed that flexibility is important, the separation between risk assessment and risk management must be retained, a distinction is needed between socially relevant and biologically relevant end-points for assessment, a social consensus as to which environmental values should be protected is needed, and scientists should communicate knowledge, not policy.

DOSE-RESPONSE ASSESSMENT
J. Bailar and J. Meyer

Discussion in this session focused first on the need to generalize the concept of dose-response assessment for ecological applications and then on the complexities that need to be addressed in practice. The group agreed immediately that for ecological assessments it is better to talk about stress-response than about dose-response relationships. Scientifically, the stress-response concept, as it applies to ecological risk assessment, is complex and involves many considerations that are absent from the usual understanding of dose-response relationships in human health risk assessment. The bulk of the session was devoted to identifying those considerations and discussing how assessments should be structured to address them.

Aspects of An Adequate Stress-Response Analysis for Ecological Risk Assessment

Selection of End Points

The group argued that end-point definition is critical for ecological stress-response assessment. Responses can be assessed at all three hierarchical levels of ecological organization: population, community, and ecosystem. Because of the inherent linkages between the levels, it is important to assess how an effect at one level can affect the other levels. No standard methods exist for making those linkages. Because empirical studies of different levels of organization usually also involve different spatial and temporal scales, the decision about which levels to study must be made *before* studies are initiated.

Final end points must be expressed as measurable characteristics, such as minimal sustainable population or maximal damage that permits the continued viability of a complex ecosystem. Both structural end points and functional end points should be considered. Structural end points include descriptive characteristics of an ecosystem, such as abundance, species composition, and trophic structure. Functional end points include energy/material flows and other transformation processes (i.e., what the organisms do, as distinct from what they are). The choice of end points must be responsive to both technical and policy concerns, including the following:

- Values (what do we really care about?),

- Measurability (can we get the data we need to do the assessment?);
- Correlation (there might be little value in studying an end point that is highly correlated with one already selected);
- Policy relevance (can the end point be linked to feasible policy options?);
- Tracking and enforcement (can future efforts tell whether the management actions based on risk assessment have been effective?).

Many ecological risk assessments necessarily deal with complex systems that offer an abundance of possible end points for study, and selection of one or a few of them for the intense effort required in a full-scale risk assessment is likely to be time-consuming and expensive—perhaps as long and expensive as the risk assessment itself.

As a strategy for selecting end points, the group consensus favored starting with a broad focus and then narrowing to the appropriate level of detail to define the design of the assessment. Taking an initially broad approach prevents missing the broader implications of hazard and stress. Institutional forms of risk assessment, such as premanufacture reviews, are so routinized that the level of organization (e.g., population) is predetermined. For noninstitutional applications, the ability to quantify will probably dictate the level of organization.

Consideration of Nonlinearities
And Discontinuities

Nonlinearities and discontinuities are likely in the response of ecological systems to stress. The group consensus was that the likelihood of observing a threshold or mean-threshold in the stress-response function increases with system complexity. Because thresholds are common in ecological systems, goals of stress-response analysis should include identification of degrees of stress at which thresholds occur and estimation of the upper ends of the threshold ranges.

The slope of the stress-response curves might be steeper as the scale of organization increases—and might approach a step function for communities and ecosystems. Therefore, the assessor needs to be sensitive to the probabilities of catastrophic changes that have few analogues at lower levels of organization and, consequently, use a greater margin of safety. Work is needed to understand the mechanisms of the response that occur at the threshold in the stress-response function.

Expression of Uncertainty

The functional expression of the stress-response relationship is stochastic and distributional; the assessor must consider extremes and discontinuities, not just central tendencies. Assessments should recognize the natural variability in systems, and conclusions should be accompanied by a description of uncertainty and probability.

Understanding the Stressor

Qualitative and quantitative aspects of the stressors should be clearly articulated without bias with respect to desirability of outcome. The effect of other anthropogenic or natural stressors should be included in the analysis, because most ecological systems are affected by multiple stresses. For example, assessments of ecological risks of chemicals could increase reliance on field experiments in which test organisms are exposed to a suite of compounds and a range of natural conditions (this approach is already being widely used to set water-quality criteria). One might also use a stressor classification to locate sensitive systems and sensitive components (e.g., species). Such a classification could include the components of the system potentially affected by the stressor, the pathway(s) of movement of the stressor, and the capacity of the affected component(s) to recover. Assessors should consider developing a matrix that considers the analytical method used to quantify stress versus class of stressor.

A good understanding of mechanisms of action can substantially improve understanding of stress-response relationships. Knowledge of mechanisms is not, however, a prerequisite for a useful risk assessment. Before a theory of mechanisms is used in a risk assessment, it must be validated in a realistic and comprehensive fashion.

Understanding the Response

Stressors have both direct and indirect effects, and both should be incorporated in the stress-response analysis. Timing of the stress (e.g., seasonality) can be critical in determining the response. A given stressor can have system-specific responses that vary geographically and with time.

Other Considerations

For the near term, some of the best data for assessing stress-response relationships in ecological systems are published empirical data on naturally or intentionally manipulated systems. For example, many whole-lake experiments have been conducted in which nutrient loading, acid deposition, and food-web structure have been manipulated (Schindler et al., 1985; Carpenter, 1989). Temporal and spatial scales of analysis should be appropriate to the stress and to the responses. For a point-source chemical release, the appropriate scale might be relatively small and short-term, depending on the dispersion pattern and degradation rate of the chemical and the life histories of the potentially affected organisms. For contamination or habitat change that affects large areas and threatens extinction of species, regional or global assessments that cover decades or centuries are appropriate.

Additions to the 1983 Paradigm Needed for Ecological Risk Assessment

The paradigm should include consideration of the resiliency of the ecological systems in question and the time to recovery. Both would vary with the type of system, geographic location, structure or process of interest, and type of stress.

The paradigm should discuss the choice of end point(s), including the relation of an end point to other major technical and societal concerns.

Application of Stress-Response
Analysis In Case Studies

The group found that some form of stress-response analysis was used in nearly all the case studies, most obviously in the chemical-related assessments. Both laboratory-derived and field-derived stress-response information was used in the TBT studies. The agricultural-chemical case study used various kinds of stress-response information, from biochemical studies to field experiments, although all focused on individual species populations. Consideration of multiple stressors would be useful. The TCDD modeling studies were driven by human health considerations and, therefore, used individual-level stress-response information.

In the spotted owl case, the stress-response relationship applies to habitat loss as the stressor and population viability as the response. In the case of harvesting, fishing is clearly the stressor, and yield and future recruitment are the end points. The latter study in particular was a good example of potential discontinuities in stress-response relationships, in that the Georges Bank haddock population, once depressed by overharvesting, did not recover after fishing pressure was relaxed. It is not clear whether the concept of stress-response relationships applies to species introductions.

Modeling Needs for Stress-Response Relationships

The group agreed that models are needed to deal with changes of state (e.g., shifts from bicarbonate to aluminum buffering), to incorporate multiple nonlinearities and discontinuities in multispecies systems, to extrapolate across ecological levels of organization (e.g., to assess the ecosystem consequences of a loss of a population or the population consequences of a loss in ecosystem function), and to make use of knowledge about synergies.

EXPOSURE ASSESSMENT
B. Leaderer and D. Porcella

The exposure-assessment group agreed that, for applicability to ecological problems, the definition of *exposure assessment* should be generalized to accommodate both nonchemical and chemical stressors. The following definition was proposed: assessment of the extent and nature of the stressor and its co-occurrence with the target. Stressors can be physical, chemical, or biological. Examples of physical stresses are habitat loss, thermal loadings, and UV radiation. Chemical stressors include toxicants and nutrients. Biological stressors include species introductions and pest organisms.

Targets for exposure assessment can be at any level of biological organization, from individual organisms to ecosystems and the biosphere. Exposure assessment can involve direct measurement, indicators of exposure, and modeling. *Extent* refers to the magnitude and spatiotemporal distribution of the stressor. *Nature* refers to the characteristics peculiar to the stressor (e.g., physical and chemical properties of a chemical contaminant).

Methods of Measuring Stressors for Ecological Exposure Assessment

The group identified a wide range of methods applicable to measuring ecological exposures. The most obvious methods are the same kinds of direct and indirect methods used in human exposure assessment, including measurements of environmental contaminant concentrations in media to which organisms are exposed, measurements of uptake or body burden, and measurements of biochemical markers correlated with contaminant exposure.

Larger-scale tools for exposure assessment include remote sensing (habitat, productivity, and albedo) and aerial and ground-based mapping. Those methods are especially appropriate for such assessments as in the spotted owl study, in which habitat change, rather than a contaminant, is the stressor. Some participants suggested that ecosystem characteristics (measures of structure and function) can be used to quantify exposure.

There was a consensus that modeling for retrospective and prospective analysis will likely play a more important role in ecological exposure assessment than it normally does in human-health risk assessment. Direct measurements with personal monitors or tissue-fluid analysis, the preferred methods of human exposure assessment, usually are not feasible or are prohibitively expensive in ecological assessments. Modeling was at least an underlying concept in all the case studies.

Test of the Definition

The group tested its proposed definition by attempting to fit it to the case studies and the 13 issues addressed by the EPA Science Advisory Board (SAB) Relative Risk Reduction Project (EPA, 1990). The group concluded that the new definition fit all six case studies, but that the definition provided in the Red Book fit only the three chemical case studies. Similarly, the new definition fit all 13 of the issues addressed by SAB, but the Red Book definition fit only about half. Two of the six case studies (on species introductions and harvesting) were related to issues *not* addressed by SAB.

RISK CHARACTERIZATION
G. W. Suter II and W. A. Farland

This group first developed a definition of risk characterization for ecological assessment and then applied the definition to the six case studies. As in health risk assessment, the principal objectives of risk characterization are to integrate information on exposure and effects and organize the results for presentation to risk managers, stake-holders, and the public.

Definition of Risk Characterization

The group determined that integration of exposure and exposure-response assessments is a complex process that requires a great deal of expert judgment. For the relatively straightforward case of predictive assessment of risks associated with chemicals, ecological effects result from exposure to contaminants in ambient media, such as air, water, and sediment. These effects are functions of both the magnitude and duration of the exposure. Risk characterization requires that the assessor identify the dimensions of exposure and effects that are relevant to the risk estimate. The assessor must determine the relative positions of the expected exposure and effects in the (perhaps multidimensional) exposure-response space and then estimate the probability that the exposure exceeds some criterion of effects.

The above description applies well to predictive risk assessments of chemicals; however, other integration approaches might be more appropriate to other types of assessments. For example, when epidemiological methods are used to assess risks associated with apparent environmental damage (decline of a population, decline of forest stands, etc.) risk characterization would include evaluation of the strength of association, the plausibility of causation (given information on mechanisms), and the extent and magnitude of the observed effects.

Components of Risk Characterization

The group determined that characterization of risks for each end point and action should include the following components (not all must be included in presentations to all audiences):

- An estimate of effects, including severity, frequency, spatial scope, temporal scope, and probability;
- A description of the sources and magnitudes of uncertainty in the risk estimate;
- An explanation of the assumptions used and a discussion of the plausible alternatives;
- A discussion of the nature and quality of the models used (types used and existence and credibility of validation studies);
- A discussion of the nature and quality of the data (quality assurance, quality control, relevance to the site, etc.);
- Supporting lines of evidence (alternative models, different types of laboratory test data, and field monitoring that support the risk estimate);

• Conflicting lines of evidence (models and data that do not support the risk estimate);

• Explanation of the weight-of-evidence determination for cases in which conflicting lines of evidence are identified;

• Description of the context of the effects estimate, including comparisons with other anthropogenic risks and with natural spatial and temporal variability and implications for other end points and levels of ecological organization that were not formally assessed;

• Identification of management actions to improve the assessment, including research or regulatory experiments that would substantially improve the risk assessment as a basis for decision-making.

Organization and Presentation

Risk characterization as the product of a risk assessment process forms a critical link between the components of the risk assessment and the risk management process. The major function of the process of risk characterization is to communicate to a risk manager the information essential to making a decision. The importance of communicating the technical bases for risk estimates to both risk managers and the public is now generally recognized. The points that follow represent highlights of a discussion concerning how risk characterizations should be organized and presented to maximize this communication.

Not only risk managers and resource managers, but also the general scientific community, the interested public, legislative bodies, and perhaps others, should be considered as potential audiences for risk characterization. The diverse nature of this audience makes the development of a good risk characterization challenging. It was agreed that the product should be tailored to meet the needs of the expected audiences. Although it needs to contain the most important scientific information, assumptions, and uncertainties, it must not be so encyclopedic that it obscures communication of major messages. The risk characterization should be viewed as a product that will connect the science to the decision-making process. Moreover, it should not be simply transmitted with no additional involvement of the risk assessor. Two-way communication should be encouraged.

The discussants agreed that the best approach to the achieving the goals described above was the development of a hierarchical information package that would contain all relevant information arrayed in increasing complexity of description. That approach would allow the characterization to be understood as deeply as deemed necessary by various clients. Discussants characterized the product as "simple words to complex graphs and charts," meaning that a good risk characterization should consist of a concise summary supported by detailed appendices.

The risk characterization should convey the nature of the uncertainties, so that clients understand that uncertainties arise from both a lack of specific knowledge and from the character of the available information (i.e., both what we do not know and what we do know contribute to our uncertainties). Finally, the characterization should convey perceived needs for information. It should clearly identify research or management practices that could reduce uncertainties or improve our ability to assess risk in the future.

Differences from and Similarities
To the 1983 Report

Discussants considered the definition of *risk characterization* presented in the 1983 report (p. 20): "Risk characterization is the process of estimating the incidence of a health effect under the various conditions of human exposure described in exposure assessment. It is performed by combining the exposure and dose-response assessments. The summary effects of the uncertainties in the preceding steps are described in this step." It was generally felt that this aspect of risk assessment was the least well-developed component of the process. Discussants were particularly concerned that the characterization not only describe the incidence of potential ecological effects, but also consider types and levels of hazard and communicate what is known about the temporal aspects of a risk. Although it was not mentioned in the 1983 report, discussants stressed the importance of having the risk characterization represent an iterative process involving collaboration between the assessors and their clients.

All those points led to a consensus that risk characterization for ecological risk assessment should not be constrained by the 1983 report. It must represent a broader perspective on the nature of potential effects.

Specific protocols for ecological risk characterization will likely be developed only through practice.

Application to the Case Studies

The ecotoxicological case studies were sufficiently similar that common lessons could be drawn from them. The population-management case studies were diverse and are discussed separately.

Lessons learned from the ecotoxicological case studies regarding risk characterization are summarized as follows:

• There was no attempt to carry out a risk characterization. None of these case studies included an actual risk assessment, and none contained attempts to convey the science in a risk perspective.

• Ecotoxicological assessments are as amenable to the development of a risk characterization as health-risk assessments.

• Case-study end points were not well characterized in general. Sentinel species were used in most cases. End points need to be put into perspective, so that the scope of the assessment and the relevance of measured versus predicted responses can be appreciated.

• The quality of the data was not made explicit in the case studies.

• Any inferences drawn on the basis of extrapolation across species or levels of organizations need to be carefully articulated, and uncertainties in them need to be explicit.

• Each case needs a statement as to how the acquired data will affect future assessment. Not all data gaps represent data needs for ecological assessment.

• Exposure-based partitioning was suggested as a means to put the scope of case assessments into perspective. It will help to determine whether a defined case represents a major or a minor route of environmental contamination or exposure.

The Georges Bank fishery study is the nearest of the population-management case studies to the conventional scheme of integrating exposure estimates (harvest) with exposure-response (fish population and community dynamics) models. Good features of this case study include development of alternative lines of evidence, acknowledgment of uncertainty, and recommendations to resource managers for management experiments. The major unresolved problem in risk characterization is communication with managers. It was clear, both from the case-study paper and from the discussion after the case-study presentation, that fishery managers are resistant to ecological risk as a decision-driver, have a short time horizon, and have difficulty in appreciating the assumptions that underlie alternative models.

As described in the case study, the northern spotted owl assessment was not formulated in terms of risks, and the decision apparently was not based on analysis of the relationship of exposure (to logging) to effects (population reduction). Literal application of the risk-characterization scheme developed by the group would require that spotted owl population characteristics be quantitatively related to habitat characteristics. Decisions concerning spotted owl management appear to have been based principally on qualitative habitat evaluation. Demographic models, such as the one presented in the case study, have been used principally as supporting lines of evidence. Uncertainty, especially concerning the link between spotted owl population dynamics and distribution patterns of old-growth forest, has not been systematically addressed.

The species-introduction case study is not a risk assessment, as defined for this workshop. There is no scale of exposure (the species is successfully introduced or not), and the effects are qualitative (the species is effective or not; it becomes a pest or not). The risk assessment is intuitive and based on expertise, rather than on explicit assumptions and models. Because the regulatory approach used by USDA is not open, it is not subject to review and scrutiny, and no attempts are made to communicate the results beyond regulatory decision-makers. There is no acknowledgment of uncertainty, even though the number of alternative hosts tested is small relative to the diversity of potentially exposed species. The case study reminded the session participants of the space-shuttle program, which relied on intuitive risk assessments until catastrophic failure occurred. To outsiders, it is not clear whether the success of USDA's species-introduction program in avoiding ecological catastrophes is due to luck or to the rigor of the evaluation program.

MODELING
R. Costanza and D. Mauriello

The group examined the case studies with regard to their use or non-use of models. Each was evaluated according to the answers to these questions:

- Were models used?
- If models were not used, would they have improved the assessment?
- If models were used, could their use have been improved?

The group also considered some general issues regarding the use of mathematical models in risk assessment and risk management.

Use of Models in the Case Studies

Tributyltin

Models were not used in the hazard-identification phase. They were used to predict the rate of leaching of TBT into the water from ships painted with antifouling paint. The decision to ban the use of the paints in Virginia was based only on hazard assessment. Such a decision might not have been made if the vulnerable organisms had not included commercially valued species.

Agricultural Chemicals

This case study described a rigorous approach to hazard identification and exposure-response assessment. Models were extensively used in determinations of the sensitivity of end- point species to pesticide exposure. The case-study paper pointed out that little basic knowledge is available on the overall ecology of agroecosystems and that this would be a fertile subject for future modeling efforts. The discussion group agreed that larger-scale models are required to deal with geographic variability and to guide future research in pesticide ecotoxicology.

Field-scale experiments could be used to validate specific risk assessment models keyed to specific pesticide-application scenarios.

Polychlorinated Biphenyls and 2,3,7,8-Tetrachlorodibenzo-*p*-Dioxin

This case study was an excellent example of the use of models to predict the environmental fate of persistent chemical contaminants. It included evaluations of uncertainty and descriptions of validation studies. In formal risk assessments, the models described are used to estimate spatiotemporal profiles of environmental contamination in sediment, water, and fish that are then compared with regulatory standards. Dose-response relationships are not used.

Species Introductions

Models are not currently part of the regulatory framework for USDA's species-introduction program. However, the author of the case study discussed research in which models of host-parasite dynamics are being used to examine the potential effectiveness of control agents proposed for introduction. This approach was viewed by the group as being analogous to defining dose-response relationships. Some attempts are being made to use models to extrapolate from test environments to other environments of interest, but models are not used for risk characterization. No methods exist for evaluating the impact of species introductions on a regional scale.

Northern Spotted Owl

The consensus of the group was that the case study was an example of hazard identification, rather than of complete risk assessment. There was no characterization of risk or uncertainty. Models might be useful for simulating the various factors that determine spotted owl population viability. A full-scale owl population model—incorporating resource availability, habitat suitability, and competitive interaction—would allow the evaluation of species-recovery options. An alternative approach would be to develop a landscape-level model that would be used to evaluate habitat management options.

Georges Bank Fishery

As described in the case study, models are used extensively to assess the status of exploited fish stocks to quantify the relationship between fishing intensity and future abundance. Risk characterizations clearly delineate the effects of alternative harvesting strategies. However, the management decision-making process was described by the case-study author as being disconnected from the scientific risk assessment exercise. The consensus of the group was that an adaptive management process, in which management itself is viewed as an experimental tool, is needed. The implementation of such an approach would require a closer connection between stock-assessment scientists and fishery managers.

General Discussion:
Models and Risk Assessment

There was general agreement that modeling should have a prominent role in risk assessment. The participants agreed that models provide the only means to perform ecological risk assessments on large physical and organizational scales. Modeling should prove especially valuable for the more complex risk assessments required in the future (e.g., for release of genetically engineered organisms). It was clear from the case studies that models are being used in some settings. However, no consistent integration of modeling into risk assessment was evident. In particular, models were not routinely used in risk characterization or in evaluation of management alternatives. The Georges Bank assessment made the most extensive use of models, but even here the results of modeling did not appear to influence decision-making.

The group advanced a number of explanations for the lack of influence of models on risk management decisions:

• Models are perceived as being too difficult to use and requiring too many data;
• Risk managers do not understand the models and have little faith in their results;
• The models are too difficult for risk assessors to use routinely;
• Models sometimes lack credibility with decision-makers, because of lack of validation or conflicting results from alternative models.

The group agreed on four possible steps to increase the use of models in ecological risk assessment:

• Development of a collaborative approach to risk assessment that includes both managers and modelers (risk assessment should be regarded as a process, not a discrete event);
• Development of models with easier-to-use front ends or expert systems to ease risk assessors into the routine use of models;
• Development of databases in tandem with models and risk assessments to provide means of validation and evaluation;
• Encouragement of quantification of uncertainty through the use of Monte Carlo methods and multiple models that incorporate alternative process formulations.

UNCERTAINTY
R. Kimerle and E. P. Smith

Evaluation of uncertainty is a critical component of all risk assessments. Sources of uncertainty include limitations in knowledge, limitations in the use of models to approximate the physical world, and limitations in the parameters that are estimated and used in models to predict risk.

Uncertainties Identified
In the Case Studies

The discussion group identified three general categories of uncertainty common to all six case studies:

• *Measurement uncertainties*, e.g., low statistical power due to insufficient observations, difficulties in making physical measurements, inappropriateness of measurements, and natural variability in organic responses to stress;
• *Conditions of observation*, e.g., spatiotemporal variability in climate and ecosystem structure, differences between natural and laboratory conditions, and differences between tested or observed species and species of interest for risk assessment;
• *Inadequacies of models*, e.g., lack of or knowledge concerning underlying mechanisms, failure to consider multiple stresses and responses, extrapolation beyond the range of observations, and instability of parameter estimates.

Implications of Uncertainty for Ecological Risk Assessment

Most of the above uncertainties affect human-health risk assessments, as well as ecological risk assessments. The consensus of the group was that knowledge-based uncertainties are often more important than uncertainties in parameter estimates. The usual statistical measures of uncertainty, p values and variance, measure only uncertainty due to random variation within the model; they do not account for uncertainties due to use of an incorrect model.

It was generally felt that the degree of uncertainty in ecological risk assessments increases with the level of biological organization. Models of ecosystem stress have higher uncertainties than models of populations and models of individual organism response. That is due in part to the increase in the number of end points available for modeling. Organism-level studies, such as single-species toxicity tests, usually have simple end points, such as survival and reproductive success. Ecosystem studies have the same end points plus additional ones that account for species interactions and measure community effects. Because of those uncertainties, ecological risk assessments still require substantial reliance on expert judgment and cannot be strictly model-based. Judgment-based approaches, such as the quotient approach to pesticide hazard assessment (described by Dr. Slimak in his plenary presentation) are often preferable to models for regulatory risk assessment.

The group noted that the degree of uncertainty that is acceptable in a study depends on the costs associated with the outcome of the risk assessment, the magnitude of expected effects, and the availability of alternatives to the hazardous agent being addressed. In the TBT study, although there were many uncertainties, once the risk to oysters was established, uncertainties about effects on other organisms were unimportant. The availability of alternatives to TBT as an antifouling agent further reduced the importance of the uncertainties.

Recommendations for Dealing
With Uncertainty

• *A discussion of uncertainty should be included in any ecological risk assessment.* Uncertainties could be discussed in the methods section of a report, and the consequences of uncertainties described in the discussion section. End-point selection is an important component of ecological risk assessment. Uncertainties about the selection of end points need to be addressed.

• *Where possible, sensitivity analysis, Monte Carlo parameter uncertainty analysis, or another approach to quantifying uncertainty should be used.* Reducible uncertainties (related to ignorance and sample size) and irreducible (stochastic) uncertainties should be clearly distinguished. Quantitative risk estimates, if presented, should be expressed in terms of distributions rather than as point estimates (especially worst-case scenarios). Power analysis or a discussion of sample size should be included in all studies involving collection of data and testing of hypotheses.

• *A continuing program of monitoring and experimental testing is needed to improve the accuracy and credibility of the process of ecological risk assessment.* There are few standards for judging the accuracy of assessments, and continuing checks need to be made to increase confidence in the process.

VALUATION
W. Desvousges and R. Johnson

The discussion leaders began by summarizing their view of the role of valuation in ecological risk assessment. Managing ecological risks requires a consistent means of comparing alternatives. Monetary values are an appropriate basis for such comparisons. Economic concerns influence several components of the risk assessment process, including hazard identification (which end points are worthy of societal concern?) and risk characterization (what are the economic implications of uncertainty?). Cost-benefit analyses are frequently a key aspect of risk management decisions.

The discussion leaders presented some methods for valuing ecological resources based on two assumptions of classical welfare economics—that societal values are sums of individual values and that people know and can express their willingness to pay (or accept compensation) for various risk policies. They then discussed some aspects of risk that influence individual decisions about willingness to pay or accept compensation:

- Amount, content, frame, and source of information;
- Decision heuristics;
- Cause of damage;
- Responsibility;
- Degree of suffering;
- Immediacy or delay of effects;
- Morbidity or mortality.

They then discussed specific issues related to determining willingness to pay for preserving ecological resources. For recreational-use values (such as fishing, hunting, and birdwatching), techniques for valuation are reasonably well established. Current research in valuation focuses on nonrecreational values. There are two principal types of such values: ecological services (sometimes called services of nature) and existence value. Ecological services are services provided by ecosystems that otherwise would have to be provided by technology. The role of wetlands in pollution abatement and flood control is a good example of an ecological service. Existence values, more vaguely defined and more controversial, are defined by people's willingness to pay for the *existence* of particular populations or ecosystems, even if they never expect to use or see them.

The discussion leaders presented a tutorial on methods used to elicit existence values with questionnaires. There was much heated discussion.

In his summary, Dr. Johnson suggested that much better communication is needed between ecologists and the public and between ecologists and economists. Ecologists need to educate the public about the importance of preservation and must learn which aspects of nature the public values most highly. Economists need help from ecologists in educating people about the interactions between natural and human systems and in understanding motives for nonuse values. Ecologists need economists to help them understand both what people care about and how intensely they care. Ecologists also need economists to communicate effectively with risk managers who face competing demands for budgetary and regulatory resources.

RISK ASSESSMENT AND THE REGULATORY PROCESS
W. Cooper and D. W. North

Risk Assessment Has Many Uses

Because there are many uses for risk assessment, many forms of risk assessment are needed. The methodological approach and the level of detail in each form might differ a great deal, depending on the purpose for which risk assessment is carried out.

For strategic planning and setting priorities, it might be appropriate to conduct risk assessments that rely on expert judgment for direct assessments of relative risk. An example of the use of such an approach is the ecological risk portion of the recent EPA Science Advisory Report on *Reducing Risk* (EPA, 1990). With the direct approach, risk is assessed on the basis of overall integrated judgment to summarize each of the risks being compared. Modeling and other analytical tools are not used directly, but they can play an important role in providing the basis for expert judgment. The result of the risk assessment is a set of risk rankings that reflect the judgment of the assessors. The assessment also includes a discussion of the reasoning underlying the assessments, with explanation for differences among the experts. Because the direct approach relies on expert judgment, rather than mathematical formalism such as model calculations or statistical analysis to reach conclusions, the direct approach can be perceived as lacking in scientific rigor. However, the direct approach can be carried out quickly and might provide extremely important guidance to nontechnical decision-makers, especially in the absence of any other form of integrated comparison among risks that are competing for scarce resources. In particular, such methods permit regulatory agencies to set priorities and research budgets in a proactive fashion. Such activities can counter the tendency to set priorities and research expenditures based on recent crises and public pressures—reaction to the pollutant of the month—rather than a comprehensive overview of competing risks.

Risk assessment is most often viewed as a quantitative process that is used to support specific risk management and resource management program decisions and policies. Among the biggest policy issues that involve ecological risk are acid deposition and global climate alteration. Neither of those was formally presented in the workshop, but participants in this work group frequently brought them up as examples of the most complex problems for ecological risk assessment. Application to problems of this scale is a massive undertaking. The six case studies were selected to be representative of major ecological issues of concern to government agencies. The case studies illustrate the complexities and uncertainties that the agencies must deal with on such issues. Participants observed that a complete risk assessment was not presented for any of the case studies. Yet, for each case study, a massive amount of information and analysis was described. At the local level, analytical resources are rarely available to deal with such a large amount of detail. But local communities and agency offices must deal with problems, such as remediation of hazardous waste sites, management of wildlife resources, and many other small-scale matters.

Risk assessment can provide scientific support to state or local agencies that are responsible for managing risk issues but lack the scientific and analytical resources of large federal agencies. Citizens groups might also have a strong interest in risk issues, but lack scientific capabilities and resources to carry out research and analysis. Risk assessment databases and monitoring efforts carried out by federal agencies to obtain baseline data can be useful to state, local, and citizens groups. Examples include the EPA-maintained IRIS database on toxic substances and the EMAP program that EPA is developing to obtain and make available data on ecological systems.

Risk assessment can provide guidance for identifying needed data and research. Such needs often become obvious when a risk assessment has been carried out, and the conclusions on risk are found to depend on critical assumptions or elements of data, for which existing uncertainties can be resolved through research.

Risk assessment can be used for "early warning"—a determination that an issue is of sufficient concern to place on the agenda so that existing policies, regulation, or legislation can be reconsidered. Advances in scientific understanding or changes in the stresses affecting an ecological system might indicate a potential for adverse changes that were not previously recognized. Recognition of the potential for adverse changes might allow these changes to be avoided through appropriate actions. Risk assessment can facilitate evaluation that permits earlier recognition and enables timely action.

In addition to situations in which timely warning of adverse changes is important, risk-based measures of ecological systems might facilitate continuing management activities to maintain, enhance, or restore the systems. Human-induced and other stresses interact in complex ways to affect ecological systems. Understanding how management policies affect the ability of an ecosystem to withstand or recover from stress will permit more effective management policies to be selected.

Different Risk Assessment Methods Are Suited to Different Risk Assessment Needs

The discussion above on the varied uses of risk assessment implies that there is not only one correct way to do risk assessment. Rather, risk assessment methods should be considered as a collection of tools from which analysts must select for the task at hand. In some cases, the tools must be developed, because the tools needed for a particular risk assessment task do not yet exist.

Risk assessment applications in similar situations might benefit from the same or very similar methodological approaches. Therefore, it will be important for public agencies and private organizations with similar needs for risk assessment to learn from each other's experiences. Both positive and negative experiences with models, databases, statistical procedures, and methods for assessing expert judgment can provide useful lessons for improving risk assessment practice. One lesson learned from ecological risk assessments is that the power of an expensive test to reject or confirm a hypothesis should be evaluated before data collection. If the data are unlikely to provide a basis for rejecting or confirming a hypothesis important to the risk assessment, then it might not be worth the expense to obtain the data.

As the volume of risk assessments grows, it will be particularly important to ensure quality and consistency. Development and use of formal guidelines, training of risk assessors, and communication of examples of good risk assessment practice will help agencies and organizations to ensure quality and consistency in their applications.

Consistency and flexibility must be balanced appropriately in the risk assessment process. Consistency motivates doing risk assessments for similar situations in a similar and predictable way. Flexibility motivates departures from a standard risk assessment approach when scientific information indicates that differences are important for the proper assessment of risk. In practice, it might be appropriate to have standard or default procedures that are used when scientific information is not sufficient to motivate a different approach, and provisions for innovative exceptions that are supported by applicable scientific information.

Risk assessment should not become too rigid. Its purpose is to summarize and communicate applicable science to meet the needs of policy makers. That task by its very nature requires flexibility and creativity, not reliance on formulas or cookbook recipes evolved from past practice.

Risk Assessors and Risk Managers
Need to Communicate

Managers responsible for ecological systems must be responsive to the public, and risk assessors should recognize that their task supports risk management. Risk assessment can help risk managers to explain the basis for their decisions to interested and potentially affected groups. Risk assessment, therefore, has an important function as communication.

As was stressed in a recent National Research Council report on risk communication (NRC, 1989), such communication should be two-way. To ensure that communication is effective and that public concerns are addressed, it is generally useful to involve the public while the risk assessment is being prepared. Interested and affected groups should be informed in advance about the risk assessment. They should have the opportunity to express their concerns and contribute information to the risk assessment process while the process is being carried out.

Issues that seem obvious to the expert scientists participating in the risk assessment might not be obvious to laypersons, but it is important for both to understand each other if there is to be effective bridging between the scientific knowledge available to the experts and the concerns of the public. As one working-group participant expressed it, "if risk assessment opens up a dialogue, then it serves an appropriate objective."

Communication between modelers, risk assessors, and managers should be mutual, iterative, timely, and flexible. Risk assessments will be valuable as support to the risk management process only if the assessments address the right problem and if the managers who are the users of the products of risk assessment understand them. One suggestion offered at the workshop is that an agency assign someone the task of being the translator, or liaison, between the group that has carried out the risk assessment and the users of the risk assessment.

Credibility is Crucial

Risk assessments will be useful to the extent that they are perceived to be effective in accomplishing a difficult task: summarizing what science can tell us about the possible consequences to an ecological system. If a risk assessment is perceived to be incomplete or biased toward a particular point of view, it will not be trusted for risk-management decision making. It is therefore essential that a risk assessment be a comprehensive and balanced summary of the applicable science.

How can comprehensiveness and balance be achieved? The recommendations on health risk issues from the 1983 report appear equally applicable to ecological risk issues:

• Regulatory agencies should take steps to establish and maintain a clear conceptual distinction between assessment of risks and the consideration of risk management alternatives; that is, the scientific findings and policy judgments embodied in risk assessments should be explicitly distinguished from the political, economic, and technical considerations that influence the design and choice of regulatory strategies.

• Before an agency decides whether a substance [ecosystem stressor] should or should not be regulated, . . . a detailed and comprehensive written risk assessment should be prepared and made publicly accessible . . .

• An agency's risk assessment should be reviewed by an independent science advisory panel before any major regulatory action or decision not to regulate.

In those recommendations, it might be appropriate to replace "regulatory" language with more general terms relevant to the broad range of decision alternatives available for the management of ecological risks. However, the principles embodied in the recommendations can be applied essentially unchanged: to promote credibility, establish and maintain the conceptual distinction between risk assessment and risk management; place risk assessments in a written, publicly accessible form; and subject them to peer review by outside scientists.

Appendix G

Contemplations on Ecological Risk Assessment

THOMAS E. LOVEJOY
Smithsonian Institution

I approach this subject with a basic concern about the environment and from a scientist's perspective. In addition, I look at it through a biological-diversity lens and with a background both in conservation biology and ecology.

What does someone concerned with the Amazon rainforest make of ecological risk assessment? It is interesting to apply the ecological risk assessment approach to the Amazon. First, the problem of Amazon deforestation is so blatant that *hazard identification* is obvious (loss of biological diversity, degraded landscapes, regional climate disruption, and greenhouse-gas production). If I think of my own research involvement with the effects of habitat fragmentation, the examination of the effect of scale (fragment size) on community structure and species richness could be considered a *dose-response* study. When the results are applied by Jim Tucker of NASA in analyzing the effects of Amazon deforestation on biological diversity, it is a form of *exposure assessment*.

It has been fascinating to look at the process of ecological risk assessment over a wide variety of case studies: from a single species (whether in terms of human health or oysters), through multispecies systems (the Georges Bank fishery) and ecosystems (what the spotted owl issue is really all about), to the scale of the entire biosphere included in the purview of Bill Cooper's committee on priority setting for EPA.

One important question that arises is, what counts as a problem? For humans narrowly, the critical end point for determining whether something is a problem is pretty clear: it is essentially a question of human health. For more complex systems there are multiple determining end points. On the one hand, there is a danger of wasting energy and time in debating what these end points might be, rather than acting on the problem. On the other hand, there is a danger of stifling constructive debate by concentrating exclusively on action. Curiously, at the planetary scale, matters seem to integrate into simplicity with clean-cut end points like stable atmospheric composition, maintenance of biological diversity, and normal levels of UVb radiation. One idea that sticks in my mind is that *what happens* is a scientific question and *what we want* is a value question. Someone else expressed the latter in another way: So what if a few robins bite the dust?

What counts as a problem—i.e., a risk—is especially complex when it comes to ecosystems. The tendency, of course, is to look primarily at ecosystem function. The rare species within an ecosystem usually play only a small role in an ecosystem's function. A focus on ecosystem *function* in risk assessments of ecosystems results in an assessment based on the few species that contribute the most to ecosystem function. A tropical forest, however, would require a different approach because a vast array of rare species constitute the bulk of the biomass and ecosystem function.

Focusing on ecosystem function tends to lead to a snapshot approach that overlooks how species have important functions at different times. I think, in this regard, of a yeast discovered by a graduate student at the Academy of Natural Sciences of Philadelphia. It is normally rare, apparently outcompeted by other organisms, because of an unusual metabolic pathway that skips over a few normal steps. When faced with rising mercury concentrations in its environment, the yeast is suddenly at an advantage, because the steps that its pathway skips are vulnerable to mercury compounds. Furthermore, the yeast is capable of reducing mercury compounds to their elemental form. The yeast population explodes, and its vacuoles fill with mercury, which is then deposited on rock surfaces. Mercury in the aquatic community is cleaned up, and the yeast becomes rare again. Examples of this sort might in fact be unusual, but the larger point is that many rare species in ecosystems can represent infrequent but recurring conditions in which those species *do* play important roles.

Indeed, a focus on ecosystem function to the exclusion of biological diversity overlooks biological diversity itself as an end point—one that represents potential resources (including genes), intellectual resources (evidence of how living systems can work), and environmental indicators. The latter brings to mind the fascinating story of TBT. One wonders whether, if people did not eat oysters, anyone would have noticed or cared that TBT at only 2 parts per trillion would cause the female dog whelk to grow a penis (imposex).

Nonetheless, I cannot help noting the difference between the oyster population today, filtering a volume of water equal to the Chesapeake Bay once a year, as opposed to once a week before the major population decline. This presents the term *keystone species* in a new, expanded light.

The question of what is sufficient evidence for action is central in the business of ecological risk assessment. For cancer producing substances, there are, for the moment, reasonably precise working definitions that use laboratory studies of what constitutes limited and sufficient evidence. In the example of TBT or Georges Bank, the evidence is more circumstantial, although common sense suggests causality. Certainly, action should not always wait for an understanding of mechanisms of causality, especially when dose-response linkage is clear. Yet at the same time, there is real value, as pointed out by Dr. Maki, in pushing ahead with research to understand causal mechanisms. That would help, for example, in evaluating substitute compounds for TBT or CFCs. In the last analysis, assessment is an iterative process, and action and policy cannot wait forever.

Another tough question is what constitutes acceptable solutions. There are, in fact, two definitions of what is acceptable: what works scientifically and what society is willing to accept. It was clear from the fisheries discussion that fishing-fleet managers resist the notion of risk as a driver of decisions. The same must be involved in the spotted owl old-growth controversy. A point made about the latter was that, although a range of alternatives were presented, only two relative extremes entered into the debate. Whether the other alternatives might have been acceptable or not, there clearly will be times when there is only one scientifically acceptable solution (habitat type being destroyed, etc.), and it is important to err on the safe side. As Robert May points out, some arguments do not have two sides—it is a figment of journalism and wishful thinking to believe that all arguments do. Some questions do not have two sides—at least not the two perceived on first examination. Sometimes the alternatives or solutions lie outside the envelope being considered. One wonders how the spotted owl/old-growth debate would have played out had the question of alternative means of making a living for the timber workers been a major part of the exercise early on.

It is in the area of solutions that the term *uncertainty* often raises its head. The term can easily be manipulated to put scientists on the defensive. Uncertainty is fundamentally part of the inherent honesty of the scientific process. Indeed, it is part of the normal way in which we discuss almost anything and is generally endemic to decision-making. Uncertainty is at least as applicable to the countervailing view of what recommended change science may put forth. We need to recognize that and not permit ourselves to be put on the defensive.

One initial talk divided ecological risk assessment into two kinds: one is essentially reactive (someone notices a problem) and involves initially retrospective and ecoepidemiological problems; the other is active and predictive. Clearly, there is a need to move toward greater emphasis on the active, but it is folly to assume that we will ever know enough to avoid surprises altogether. Clearly, too, there is a need to set priorities: it is impossible to do everything at once. There is a serious challenge, in that ecological risk assessment is vastly more complicated than society will ever understand, and at least in principle, society wants it *all* done. That puts a tremendous premium on sound risk assessment, on communication, and on priority-setting. EPA's effort to set priorities, giving great emphasis to problems of substantial spatial and time dimensions, is an extremely important exercise—essentially ecological risk assessment on a planetary scale.

Early in the workshop, I participated in a discussion of the role of science vs. the role of values in setting the agenda. A devil's advocate suggested that the role of science was to work on problems once values were set. There is, of course, some truth in that: if society wants something, science responds. But in setting those values, society must be *informed* by science. The new priorities set by EPA's Science Advisory Board are only partly congruent with society's priorities, or society's historical priorities as reflected by EPA's budget. Obviously, there is a long way to go. From the tidewater to the ozone layer, ecological risk assessment has a vital role to play.

Appendix H

Workshop Summary

The workshop lasted only 3 days, and it was impossible to achieve consensus on every issue. There was general agreement on the need for ecological risk assessment to be broadly defined. As noted in the plenary presentations by Drs. Yosie, Lovejoy, and North, the policy needs that must be served are broad. Despite the diversity of environmental problems and the complexity of the science needed to address them, decision-makers need common frameworks for comparison and common procedures to ensure credibility.

Retrospective studies, such as those of TBT and the spotted owl controversy, which involve identification and resolution of existing problems, and predictive studies, such as those of agricultural-chemical regulation and biological control, which are aimed at preventing new problems, involve different scientific approaches and rest on different information bases. The technical issues discussed at the workshop include the following:

- Selecting among numerous possible end points at different levels of biological organization;
- Extrapolating effects from one species or level of organization to others;
- Discontinuities and nonlinear responses;
- Spatiotemporal scaling;
- Accounting for background variability;
- Evaluating both quantitative and qualitative uncertainties.

Despite the complexities, there was a clear consensus that it is feasible to talk about assessing ecological risks in a manner analogous to human health risk assessment and that most, and perhaps all, types of ecological risk assessment can be accommodated within a single conceptual framework.

There was also a consensus that the health risk assessment framework presented in the NRC's 1983 report, although useful as a point of departure, is too narrowly defined for ecological risk assessment. Its most obvious weakness is its orientation toward toxic chemicals. Ecological risk assessment must be applicable to a much broader array of stresses. The discussion groups on exposure assessment and dose-response assessment agreed, however, that the concepts of exposure and dose response could be generalized in a straightforward way to accommodate nonchemical stresses.

A clear theme running through nearly all the case studies and discussion groups was that the links between management and risk assessment are much stronger and more pervasive in ecological risk assessment than is indicated in the 1983 report. Subjects of particular importance include the role of policy, in the form of legal mandates and regulatory procedures, in defining an ecological hazard, the kinds of information to be used to assess risks, and the complexity of risk characterization in ecological risk assessment. Ecological risk assessment must include evaluations of kinds of uncertainty usually absent from health risk assessment, expression of risks in terms useful for decision-making (including economic valuations), and communication between risk assessors and risk managers, many of whom are not trained as ecologists. Several groups discussed possible modifications of the Red Book paradigm, but workshop participants as a whole were divided over whether to modify the paradigm or to develop a new one that is explicitly ecological.

A number of research-needs themes surfaced at the workshop. These are too numerous to list here and are noted in the summaries of individual discussion groups. However, we note several common themes, some of which were discussed in more than one group:

• *Extrapolation across scales.* Effects of interest to risk managers usually involve changes in populations or ecosystems. However, many stressors (including toxic chemicals and exploitation) act through direct effects on individual organisms. Alternatively, risk managers might be interested in effects of large-scale regional change over long periods (e.g., logging of old-growth forest), but individual studies are restricted to relatively short periods and small areas. Some form of modeling appears generally necessary to make these extrapolations, but few models have been used.

• *Quantitative and qualitative analysis of uncertainty.* It was amply noted that uncertainty in ecological risk assessment extends far beyond uncertainty in individual parameter values. Many of the uncertainties are related to extrapolations across scales and other kinds of qualitative gaps in knowledge. Those knowledge-based uncertainties result in many assessments being based principally on professional judgment, rather than on quantitative analysis. Evaluating the uncertainty inherent in professional judgments is as important as quantifying the uncertainty in model-based assessments.

• *Validation of predictive tools:* Needs for validation were mentioned specifically in the risk characterization, uncertainty, and modeling groups and by plenary-session speakers (Yosie and North). Validation could include both designed experiments and retrospective monitoring of the outcome of risk management decisions.

• *Expression of risks in policy-relevant terms.* This topic was debated at length in discussions of risk characterization, the regulatory process, and valuation and was mentioned in plenary-session presentations (by North, Yosie, and Slimak). Many difficulties were noted. Terms used by ecologists (such as *ecosystem, stability,* and *resilience*) are unfamiliar to decision-makers and the public, and their value is not immediately obvious. Expression in economic terms is attractive and is favored by decision-makers. However, the valuation discussion made it clear that many aspects of valuing ecological resources—especially non-use values, such as biodiversity—involve economic theories and measurement methods that themselves are highly uncertain.

Some of the above issues might never be fully resolved. However, workshop participants familiar with health risk assessment often noted that the same or similar difficulties also affect health risk assessment, and the existence of difficulties has not precluded health risk assessments. In his closing statement, Dr. Barnthouse noted that the terms of discussion about ecological risk assessment have changed. In past years, the discussion was about whether the concept of risk and the methods of quantitative risk assessment were even applicable to ecological problems. Future discussions will concern conceptual form, technical development, and implementation in specific circumstances. The reality of ecological risk assessment is now beyond dispute.

Appendix I

References for Appendixes

Abel, R., and others. 1986. Pp. 1314-1323 in Proceedings of IEEE Oceans '86 Organotin Symposium, Vol 4. Washington, D.C.

Abbruzzese, B., S.G. Liebowitz, and R. Summer. 1990. A Synoptic Approach to Wetland Designation: A Case Sudy in Washington. EPA 600/3-90-006. U.S. Environmental Protection Agency, Corvallis, Ore.

Alzieu, C. 1986. Pp. 1130–1134 in Proceedings of IEEE Oceans '86 Organotin Symposium, Vol 4. Washington, D.C.

Blunden S.J., and A. Chapman. 1982. Pp. 110-159 in Organometallic Compounds in the Environment, P.J. Craig, ed. New York: John Wiley and Sons.

Burnham, K.P., D.R. Anderson, G.C. White, C. Brownie, and K.H. Pollock. 1987. Design and Analysis Methods For Fish Survival Experiments Based On Release-Recapture. American Fisheries Society Monograph 5.

Carpenter, S.R. 1989. Replication and treatment strength in whole-lake experiments. Ecology 70:453-463.

DeBach, P. 1974. Biological Control by Natural Enemies. London: Cambridge University Press.

Endicott, D.D., W.L. Richardson, D.M. Di Toro. 1989. Lake Ontario TCDD Modeling Report. EPA Large Lakes Research Station, 9311 Groh Rd., Grosse Ile, Mich., 48138.

Endicott, D.D., W.L. Richardson, T.F. Parkerton, D.M. Di Toro. 1990. A Steady State Mass Balance and Bioaccumulation Model for

Toxic Chemicals in Lake Ontario. Report to the Lake Ontario Toxics Management Committee. EPA Large Lakes Research Station, 9311 Groh Rd., Grosse Ile, Mich., 48138.

EPA (U.S. Environmental Protection Agency). 1982. Pesticide assessment guidelines. Subdivision E - Hazard Evaluation: Wildlife and Aquatic Organisms. Office of Pesticide and Toxic Substances, U.S. Environmental Protection Agency, Washington, D.C.

EPA (U.S. Environmental Protection Agency, Science Advisory Board). 1990. Reducing Risk. EPA SAB-EC-90-021. U.S. Environmental Protection Agency, Washington, D.C.

EPA (U.S. Environmental Protection Agency). 1992a. Framework for Ecological Risk Assessment. EPA 630/R-92-001. U.S. Environmental Protection Agency, Washington, D.C.

EPA (U.S. Environmental Protection Agency). 1992b. Peer Review Workshop Report on a Framework for Ecological Risk Assessment. EPA/625-91-022. U.S. Environmental Protection Agency, Washington, D.C.

Fite, E.C., L.W. Turner, N.J. Cook, C. Stunkard, and R.M. Lee. 1988. Guidance Document for Conducting Terrestrial Field Studies. U.S. Environmental Protection Agency, Ecological Effects Branch, Hazard Evaluation Division, Office of Pesticide Programs.

Fogarty, M.J., R.K. Mayo, F.M. Serchuck, and F.P. Almeida. 1989. Trends In Aggregate Fish Biomass and Production On Georges Bank. NAFO SCR Doc. 89/81. North Atlantic Fisheries Organization Scientific Council.

Foy, C.L., D.R. Forney, and W.E. Cooley. 1983. History of weed introductions. Chapter 4 in Exotic Plant Pests and North American Agriculture, C.L. Wilson and C.L. Graham, eds. New York: Academic Press.

Franklin, A.B., J.P. Ward, R.J. Guitérrez, and G.I. Gould. 1990. Density of northern spotted owls in northwest California. J. Wildl. Mgmt. 54:1–10.

Ginzburg, L.R., L.B. Slobodkin, K. Johnson, and A.G. Bindman. 1982. Quasiextinction probabilities as a measure of impact on population growth. Risk Anal. 2:171–181.

Grosslein, M.D., R.W. Langton, and M.P. Sissenwine. 1980. Recent fluctuations in pelagic fish stocks of the northwest Atlantic, Georges Bank, region. Rapp. P.-v Reun. Cons. Int. Explor. Mer

177:374-404.

Hooper, M.J., P. Detrich, C. Weisskopf, and B.W. Wilson. 1989. Organophosphate exposure in hawks inhabiting orchards during winter dormant-spraying. Bull. Environ. Contam. Toxicol. 42:651-659.

Howarth, F.G. 1991. Environmental impacts of classical biological control. Annu. Rev. Entomol. 36:485-509.

Huggett, R.J., M.A. Unger, P.F. Seligman, and A.O. Valkirs. 1992. The marine biocide tributyltin. Environ. Sci. Technol. 26:232-237.

Kendall, R.J. 1992. Farming with agrochemicals: The response of wildlife. Environ. Sci. Technol. 26:239-245.

Kendall, R.J., L.W. Brewer, T.E. Lacher, B.T. Marden, and M.L. Whitten. 1989. The Use of Starling Nest Boxes for Field Reproductive Studies: Provisional Guidance Document and Technical Support Document. EPA/600/8-89/056. Available as NTIS Publication No. PB89 195 028/AS.

Kimerle, R.A., W.E. Gledhill, and G.J. Levinskas. 1978. Environmental safety assessment of the new materials. In Estimating the Hazard of Chemical Substances to Aquatic Life, K. Dickson, A. Maki, and J. Cairns, eds. ASTM STP 657. American Society for Testing and Materials.

NAPAP (National Acidic Precipitation Assessment Program). 1989. Acidic Deposition: State of the Science and Technology, Vol. II. Aquatic Processes and Effects. Washington, D.C.: U.S. Government Printing Office. December.

NRC (National Research Council). 1983. Risk Assessment in the Federal Government: Managing the Process. Washington, D.C.: National Academy Press. 191 pp.

NRC (National Research Council). 1986. Ecological Knowledge and Environmental Problem-Solving: Concepts and Case Studies. Washington, D.C.: National Academy Press. 388 pp.

NRC (National Research Council). 1989. Improving Risk Communication. Washington, D.C.: National Academy Press. 332 pp.

Peterman, R.M., and M.J. Bradford. 1987. Statistical power of trends in fish abundance. Canad. J. Fish. Aquatic Sci. 44:1879-1889.

Pimentel, D., C. Glenister, S. Fast, and D. Gallahan. 1984. Environmental risks of biological pest controls. Oikos 42:283-290.

Sailer, R.I. 1983. History of insect introductions. Ch. 2 in Exotic Plant Pests and North American Agriculture. New York: Academic

Press.

Salwasser, H. 1986. Conserving a regional spotted owl population. Pp. 227-247 in Ecological Knowledge and Environmental Problem Solving. Washington, D.C.: National Academy Press.

Schindler, D.W., K.H. Mills, D.F. Malley, D.L. Findlay, J.A. Shearer, I.J. Davies, M.A. Turner, G.A. Lindsey, and D.R. Cruikshank. 1985. Long-term ecosystem stress: The effecxts of years of experimental acidification on a small lake. Science 228:1395-1401.

Schweer, G., and P. Jennings. 1990. Integrated risk assessment for dioxins and furans from chlorine bleaching in pulp and paper mills. EPA 560/5-90-011. U.S. Environmental Protection Agency, Washington, D.C.

SETAC (Society of Environmental Toxicology and Chemistry). 1987. Research Priorities in Environmental Risk Assessment. Society of Environmental Toxicology and Chemistry, Pensacola, Fla.

Sissenwine, M.P., and J.G. Shepherd. 1987. An alternative perspective on recruitment overfishing and biological reference points. Can. J. Fish. Aquat. Sci. 44:913-918.

Thomann, R.V., and D.M. Di Toro. 1983. Physico-chemical model of toxic substances in the Great Lakes. J. Great Lakes Res. 9(4):474-496.

Thomann, R.V. and J.P. Connolly. 1983. A model of PCB in the Lake Michigan lake trout food chain. Environ. Sci. Tech. 18(2):65-71.

Appendix J

Workshop Program

February 26 - March 1, 1991

Airlie Foundation
Warrenton, VA

TUESDAY, FEBRUARY 26, 1991

4:00   WELCOME
Bernard Goldstein, Committee Chairman, Robert Wood
Johnson Medical School

4:15   DISCUSSION OF PURPOSE
Lawrence Barnthouse, Workshop Chairman, Oak Ridge
National Laboratory

KEYNOTE SPEECHES

4:30   **Broad Policy View**, Terry Yosie, American Petroleum
Institute

5:00   **Relationship of Workshop to NRC's 1983 Risk
Assessment Study**
Warner North, Decision Focus, Inc.

5:30   **U.S. EPA Activities in Ecological Risk Assessment**
           Michael Slimak, ORD, U.S. EPA

WEDNESDAY, FEBRUARY 27, 1991

CASE STUDIES

CHEMICAL STRESSORS

8:30   **Tributyl tin**
           Robert Huggett, Virginia Institute of Marine Science
8:55   Discussants: Lawrence Barnthouse, Oak Ridge National
                         Laboratory
9:05   Peter Seligman, Naval Ocean Systems Center
9:15   General Discussion

9:30   **Agrochemicals**
           Ronald Kendall, Clemson University
9:55   Discussants: Bill Williams, Ecological Planning and
                         Toxicology, Inc.
10:05   James Gagne, American Cyanamid Company
10:15   General Discussion

10:30   Break

11:00   **PCBs and TCDD**
           Dominic DiToro, Manhattan College
11:25   Discussants: Dennis Paustenbach, McLaren/Hart
11:35   Larry Burns, U.S. EPA, Athens, GA
11:45   General Discussion

noon   Lunch

OTHER STRESSORS

1:30   **Habitat loss** (spotted owl)
           David Anderson, U.S. Fish & Wildlife Service

1:55	Discussants: Orie Loucks, Miami University, OH
2:05	Mary Kentula, U.S. EPA, Corvallis, OR
2:15	General Discussion

2:30 **Introduction of species**
Ray Carruthers, USDA, Cornell University
2:55 Discussants: David Policansky, National Research Council
3:05 James Carlton, Williams College
3:15 General Discussion

3:30 Break

4:00 **Harvesting** (George's Bank Fisheries)
Andrew Rosenberg, National Marine Fisheries Service
4:25 Discussants: J. Larry Ludke, U.S. Fish and Wildlife
Service
4:35 Randall Peterman, Simon Fraser University
4:45 General Discussion

5:00 Adjourn

THURSDAY, FEBRUARY 28, 1991

BREAKOUT SESSIONS

7:30 Breakfast

8:30 RELEVANCE OF NRC 1983 RED BOOK PARADIGM TO
ECOLOGICAL RISK ASSESSMENT

Hazard identification
Alan Maki, Exxon Corporation
Dorothy Patton, Risk Assessment Forum, U.S. EPA

Dose-response assessment
John Bailar, McGill University School of Medicine
Judy Meyer, University of Georgia

Exposure assessment
Brian Leaderer, Yale University School of Medicine
Don Porcella, Electric Power Research Institute

Risk characterization
William Farland, OHEA, U.S. EPA
Glenn Suter, Oak Ridge National Laboratory

10:00   Break

10:30   Reconvene

12:30   Lunch

1:30   ANALYTICAL ISSUES IN ECOLOGICAL RISK
       ASSESSMENT

Modeling
Robert Costanza, University of Maryland
David Mauriello, U.S. EPA

Uncertainty
Richard Kimerle, Monsanto Company
Eric Smith, Virginia Polytechnic Institute and State
    University

Valuation
William Desvousges, Research Triangle Institute
F. Reed Johnson, U.S. Naval Academy

Relationship of risk assessment to regulatory process
Warner North, Decision Focus, Inc.
William Cooper, Michigan State University

3:30   Break

4:00   Reconvene

6:00   Adjourn

6:30 Dinner
Speaker
Erich W. Bretthauer, Assistant Administrator for Research
and Development, U.S. EPA

FRIDAY, MARCH 1, 1991

CONCLUDING PLENARY SESSION

7:30 Breakfast

RELEVANCE OF NRC 1983 RED BOOK PARADIGM TO
ECOLOGICAL RISK ASSESSMENT

8:00 CONTEMPLATIONS ON ECOLOGICAL RISK
ASSESSMENT
Thomas Lovejoy, Smithsonian Institution

8:30 **Hazard identification**, Maki and Patton

8:45 **Dose-response assessment**, Bailar and Meyer

9:00 **Exposure assessment**, Leaderer and Porcella

9:15 **Risk characterization**, Farland and Suter

9:30 **Discussion**, North

10:00 Break

ANALYTICAL ISSUES IN ECOLOGICAL RISK
ASSESSMENT

10:15 **Modeling**, Costanza and Mauriello

10:30 **Uncertainty**, Kimerle and Smith

10:45 **Valuation**, Desvousges and Johnson

11:00   **Relationship of risk assessment to regulatory process,**
         North and Cooper

11:15   **Discussion,** Barnthouse

11:45   **Conclusion,** Barnthouse

noon   Lunch/Adjourn